JOURNAL OF ICT STANDARDIZATION

Volume 1, No. 2 (November 2013)

Special Issue on

GISFI Standardization Activities

Guest Editor:

Balamuralidhar P
TCS Innovation Labs, Bangalore, India

JOURNAL OF ICT STANDARDIZATION

Chairperson: Ramjee Prasad, CTIF, Aalborg University, Denmark
Editor-in-Chief: Anand R. Prasad, NEC, Japan
Advisors: Bilel Jamoussi, ITU, Switzerland
Jesper Jerlang, Dansk Standard, Denmark

Editorial Board
Kiritkumar Lathia, Independent ICT Consultant, UK
Hermann Brandt, ETSI, France
Kohei Satoh, ARIB, Japan
Sunghyun Choi, Seoul National University, South Korea
Ashutosh Dutta, AT&T, USA
Alf Zugenmaier, University of Applied Sciences Munich, Germany
Julien Laganier, Juniper Networks, USA
John Buford, Avaya, USA
Monique Morrow, Cisco, Switzerland
Vijay K. Gurbani, Alcatel Lucent, USA
Henk J. de Vries, Rotterdam School of Management, Erasmus University, The
Netherlands
Yoichi Maeda, TTC Japan
Debabrata Das, IIIT-Bangalore, India
Signe Annette Bøgh, Dansk Standard, Denmark
Rajarathnam Chandramouli, Stevens Institute of Technology, USA

Objectives:

- Bring papers on de-jure as well as de-facto standards to the readers
- Cover pre-development, including technologies with potential of becoming a standard, as well as developed / deployed standards
- Publish on-going work with potential of becoming a standard technology
- Publish papers giving explanation of standardization process
- Publish tutorial type papers giving new comers a understanding of standardization

Aims & Scope

- Aim:
  - The aim of this journal is to publish standardized as well as related work making "standards" accessible to a wide public – from practitioners to new comers.
  - The journal aims at publishing in-depth as well as overview work including papers discussing standardization process and those helping new comers to understand how standards work.

- Scope:
  - Bring up-to-date information regarding standardization in the field of Information and Communication Technology (ICT) covering all protocol layers and technologies in the field

JOURNAL OF ICT STANDARDIZATION

Volume 1, No. 2 (November 2013)

Published, sold and distributed by:
River Publishers
PO box 1657
Algade 42
9000 Aalborg
Denmark
Tel.: +4536953197

www.riverpublishers.com

Journal of ICT Standardization is published three times a year. Publication programme, 2013–2014: Volume 1 (3 issues)

ISSN: 2245-800X

Editorial: Special Issue on Global ICT Standardisation Forum for India (GISFI)

Balamuralidhar P

TCS Innovation Labs, Bangalore

It is my great pleasure to introduce this special issue focusing on the contributions from Global ICT Standardisation Forum for India (GISFI). The Global ICT Standardization Forum for India (GISFI),has been making noteworthy contributions towards Indian standardization in the area of Information and Communication Technologies (ICT) as applicable to areas, such as energy, healthcare, security and green technologies. The effort of GISFI to create a synergy and coherence for the efforts from India in the world standardization process has been strengthened with the collaboration of various international standards bodies. GISFI has put special efforts to bring academia also into the standardization stream along with industry from diverse strata of the society. Further, there is afocus on strengthening the ties among leading and emerging scholars and institutions in India and the world around; to develop and cultivate a research and development agenda for the ICT field. The working groups organized in GISFI have attracted participation from academia, business, and Government/policy-making entities.

In this special issue we bring forth papers reporting the guiding thoughts and key contributions from each of the work groups authored by the leading contributors. Major contributions are from the following work groups.

- Spectrum Issues
- Security and Privacy
- Cloud and Service Oriented Networks (CSeON)
- Internet of Things (IoT)
- Green ICT (GICT)
- Future Radio Networks
- Special Interest Group

This special issue opens up with a vision paper from the founding chairman of GISFI on the potentially high impact of emerging technologies for 5G wireless communications. Interoperability, ubiquity and dynamism are identified to be the key features of 5G networks. Further the current activities and future directions for a special interest group for 5G in GISFI are outlined.

Wireless spectrum is an important national resource and it faces pressure due to the extreme user densities and diversity of technologies deployed. Efficient spectrum management is in the focus and there are two papers included on this. The first one is on Spectrum Challenges for Modern Services summarizing the deliberations of Spectrum Work group on improving spectrum utilization through spectrum sharing. In the second paper on "Spectrum Trading in India and 5G" authors discuss a comprehensive view on spectrum trading in the Indian context.

Today telecommunications has become the lifeline of human society and supporting structure to economy. Safety and Security is of paramount importance. Moreover the privacy of information exchange needs to be respected. The works of security and privacy workgroup in this space is presented in 'Security and Privacy'.

The 'National Telecom Policy 2012' has identified Cloud and Service Oriented Networks as an integral part of the convergence strategy laid out by Government of India. Understanding the infrastructure requirements and identifying gaps are of national interest. In the paper "Cloud and Service Oriented Networks (CSeON)" the authors summarize the contributions and achievements of the CSeON work group on this topic.

Imparting energy efficiency in systems through the adoption of wide-ranging spectrum of environmentally friendly technologies that power the connected information infrastructure is another area of focus. In the paper "Green Information and Communication Technology Standards Development: An Indian Perspective" the authors present the contributions and collaborative activities GICT Working Group (WG) in this space.The contribution on metrics and methods of measurement for energy efficiency of telecommunication equipments is very relevant from Indian context.

Tracking and managing carbon emission in telecommunication systems is an important activity under Green ICT. A techno-managerial perspective to estimate carbon emission in Indian Telecom networks is brought up in the paper "Carbon Emission Estimation & Reduction in Indian Telecom Operations". Further the author goes forward to propose a solution framework to address this objective.

Another important disruptive influence on ICT and its applications is from the vision of 'Internet of Things' (IoT). Its potential impact on different facets of human life has been widely recognized. The paper titled 'Lightweight platforms for Internet of Things' presents the works of IoT workgroup in this topic. The contributions from the IoT workgroup are majorly towards identifying relevant applications from the context of India and conceptualizing a unified service platform for supporting them. In a second paper on this topic the author discusses interoperability aspects of some of the key IoT standards and presents how that study has contributed to shapeup the platform architecture in GISFI IoT WG.

I would like to thank all the authors of this special issue for their contributions. Also grateful to all the reviewers for their timely expert reviews.I am sure that the dissemination of some of the selected works of GISFI work groups through this special issue will encourage more ICT experts from academia, businesses and administration to contribute and address some of the national challenges though ICT standardisation.

About the Editor

Balamuralidhar P is a Principal Scientist and Head of TCS Innovation Lab at Tata Consultancy Services Ltd (TCS), Bangalore. He has obtained Bachelor of Technology from Kerala University and Master of Technology (MTech) from IIT Kanpur. His PhD is from Aalborg University, Denmark in the area of Cognitive Wireless Networks. Major areas of current research include different aspects of Cyber Physical Systems, Sensor Informatics and Networked Embedded Systems. Before TCS his research careers were with Society for Applied Microwave Electronics Engineering & Research (SAMEER) Mumbai and Sasken Communications Ltd Bangalore.

He has over 25 years of research and development experience in Signal Processing, Embedded Systems and Wireless Communications. He has over 60 publications in various international journals and conferences and over 20 patent applications. Balamuralidhar was the leading TCS participation in two EU FP6 research consortium projects namely My Adaptive Global NET (MAGNET) and End to End Reconfigurability (E2R) in the area of next generation wireless communications. He is also contributing to TCS participation in National bodies like Broadband Wireless Consortium India (BWCI), Global ICT Standards for India (GISFI). In GISFI he is chairing the Internet of Things Workgroup.

Global ICT Standardisation Forum for India (GISFI) and 5G Standardization

Ramjee Prasad

Founding Chairman GISFI & Director of Center for TeleInfrastruktur (CTIF), Aalborg University, Denmark; email: ramjee.prasad@gisfi.org, prasad@es.aau.dk

Received July 2013; Accepted August 2013

Abstract

This paper covers a selected set of potentially high impact emerging technologies for 5G wireless communication as well as their current state of maturity and scenarios for the future. 'Interoperable', 'ubiquitous' and 'dynamic' are key words describing 5G networks and applications. The paper focuses on what are the necessary critical innovations, also undertaken within standardization and how to explore these as well as what are the expected challenges. It further identifies future directions for 5G and gives a vision of such communication system.

Keywords: Wireless innovative system for dynamically operating mega-communications (WISDOM), 5G, standardization.

1 Introduction

The International Telecommunication Union (ITU) [1] defines the standardization gap as disparities in the ability of developing countries, relative to developed ones, to access, implement, contribute to and influence international ICT standards and recommendations. In addition, for making international standards universally effective, it is important to identify and incorporate

Journal of ICT Standardization, Vol. 1, 123–136.
doi:10.13052/jicts2245-800X.12a1

solutions for the country-specific scenarios. The Global ICT Standardisation Forum for India (GISFI) [2] was founded in 2009 as an Indian standardization body active in the area of Information and Communication Technologies (ICT) and related application areas, such as energy, telemedicine, wireless robotics, biotechnology in an effort to create a new coherence and strengthen the role of India in the world standardization process by mapping the achievements in ICT in India to the global standardization trends.

GISFI bases its work on the Indian Governmental policy and at the same time is cooperating with international standardization organizations to create internationally viable standards solutions that reflect the specifics of the Indian ICT scenario.

The Government of India direction for the Telecom growth in the country as specified in [3] states the Indian Telecom and ICT needs as related to cost-effective, energy-efficient and suitable for the Indian environment and user-specifics, technical products and solutions.

Figure 1 shows the standards evolution since the beginning of the wireless communication era till date. During the pioneer era, much fundamental research and development in the field of wireless communications took place. The cellular era started in 1980s with the launch of the analog cellular systems, followed by the digital wireless communication systems. Despite the large variety of existing communication systems, each development has been motivated by the same goal: to provide universal service facilities to the users. In the context of the Indian scenario, where despite the overall very high wireless penetration (862 million in January 2013 [4]), rural user density remains low due to the high deployment costs making services unaffordable by those users, the ubiquitous service provision is even more challenging.

Typically, 70% of the capital cost of cellular networks is in the access network as opposed to the backbone, which means that these networks depend on a certain user density for profitability. Hence, urban areas tend to be covered by multiple carriers, while rural areas typically have one or none.

When aiming to evolve the growth of the Tree of Standards (see Figure 1) to reach the 5G top, GISFI looks at a technology concept of a system, namely the Wireless Innovative System for Dynamically Operating Mega-Communications (WISDOM) [5, 6, 7] with the following key features:

- Combines established, competitive cellular standards with a promising frequency spectrum and novel enabling technologies;
- Reduced coverage, electricity and operational expenditure (OPEX) costs;
- Offers scalable and flexible technology options.

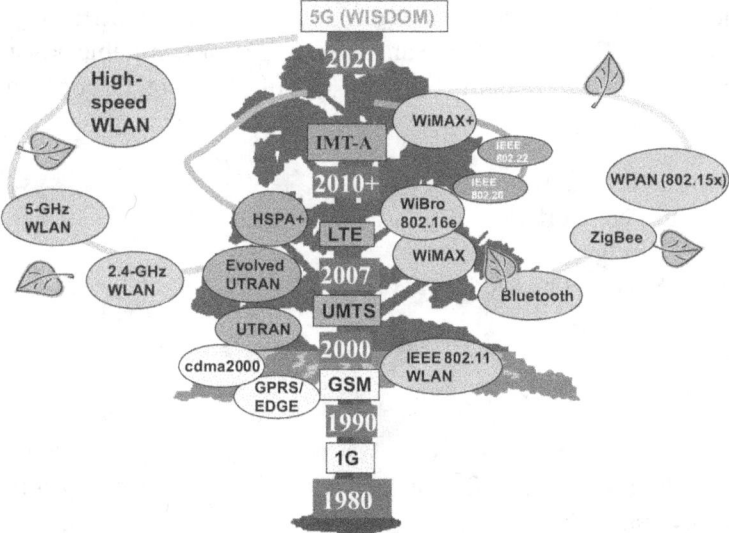

Figure 1 The tree of standards.

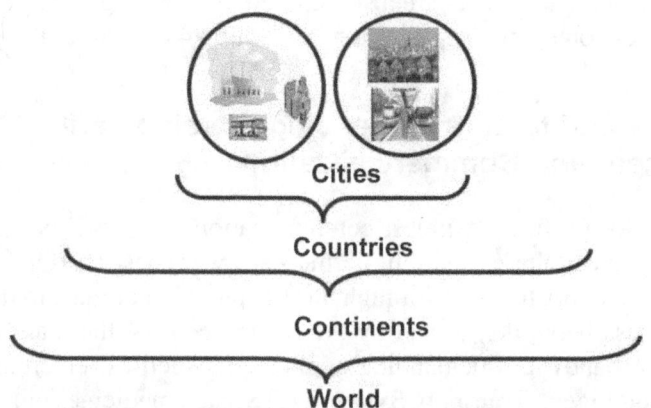

Figure 2 Global connectivity through WISDOM [5, 6, 7].

GISFI offers the vision of a future universally deployable wireless communication system that would make the pillar for enabling smart teleinfrastructures, which will offer services and applications of data rate more than tera/bit/second (Tbps), with a coverage extending from a city region, to a country, the continents, and the world to enable human-centric megacommunication applications. The WISDOM concept is shown in Figure 2.

To achieve this vision, GISFI brings different major stakeholders together into its six Working Groups to define and specify the most suitable advanced technologies to lead to enabling 5G. The GISFI Working Groups are as follows [8]:

- Security and Privacy;
- Future Radio Networks;
- Internet of Things;
- Cloud and Service Oriented Networks;
- Green ICT;
- Spectrum;
- Special Interest Group.

This paper defines the steps undertaken within GISFI towards 5G standardization including the specifics of the Indian ICT scenario and offers a vision of a universally deployable 5G system.

The paper is organised as follows. Section 2 defines the scenarios and factors influencing the development, deployment and commercialization of 5G. The main research enablers are also outlined. Section 3 gives a vision of 5G and identifies some open research issues. Section 4 concludes the paper.

2 Scenarios and factors Influencing Development, Deployment and Commercialization of 5G

Possible scenarios for future wireless communications (e.g., 5G) based on an analysis of key uncertainties and long-term trends were described in [7].

It is envisioned that the need for high throughput will continue to increase in future wireless networks, driven by the rising needs of the mass market in the fields of bandwidth demanding applications such as: entertainment, multimedia, Intelligent Transport Systems (ITS), telemedicine, emergency and safety/security applications.

Futuristic applications such as: 3-dimensional (3D) Internet, virtual and augmented reality that combines data for all senses (audio, visual, haptic, digital scent), telehaptic applications like planet or deep sea exploration, networked virtual reality (e.g., video streaming from social networks - each user streams its own reality), and telepresence (e.g., immersive environments with applications in both the commercial and military fields), can push the demand for real-time, symmetric, wireless ubiquitous connectivity to an individual with a data rate well in excess of 250 Mbps – a rate that far exceeds the capacity offered by current and emerging next generation wireless technologies and networking infrastructures.

At the same time, tremendous growth of the number of networked devices leading to the so-called trillion device network requires wireless solutions which are cost sensitive but also able to support mobility, since a large fraction of the devices will be associated with people and robots.

People's needs are changing too, including the needs to communicate in a very dynamic environment. In parallel, the demand for an efficient communication service, closer to the users, is rising. An efficient communication system is one able to follow the user wherever the user is, and able to adapt its traffic capabilities to the dynamic user environment.

In a first and most obvious direction, the development of future wireless communications will be determined by a very high public demand for wireless solutions and breakthroughs. A key challenge for reaching this scenario will be depending on whether technologies would deliver to match such high level of user expectations.

2.1 Small Cell Deployment

Small cell deployment has been identified as a means to dynamically bring the communication services closer to the user [9, 10]. Small cells can be deployed where there is a temporary need, and areable to cover the traffic needs, which arise in an area for a limited time or may follow the users in their movement. And in many cases small cells are not owned by the network provider, but are directly in the ownership of the user or are part of the communications interface offered by a third party (e.g., a transport company). A user will not consider a small cell close if this small cell is not cost-efficient. The ability to be cost-efficient is coming from the capability to adapt to the user needs, to offer connectivity when and where the user wants it, to not waste expensive energy and use only the necessary resources (e.g. radio spectrum, capacity, etc). It must be cost-efficient by itself and must be used in a cost-efficient manner. Because the users' needs and habits vary from one moment to another, from a particular user to another, small cells must support heterogeneity and reconfigurability.

The main focus of the recent technical work with in GISFI (WG Future Radio Networks [8]) has focused on how to enable additional capacity, improved coverage, and energy/cost-efficient networks by the design, analysis, development and evaluation of procedures for fast network deployment and dynamic reconfiguration of small cells in various scenarios, also relevant to the Indian specifics. The processes needed to deploy small cells economically (how to deploy, where to deploy, how to deal with the increased number of small cell sites, etc.) are also being studied in order to propose new, economically viable mechanisms.

2.2 Security

The new age of wireless envisioned in the 1990s has finally come to pass. Remarkable achievements in wireless and other areas of telecommunications (e.g., ICT) have ushered in the new age of ubiquitous mobile communications. For such scenario to emerge successfully, wireless communication systems must deliver technological solutions that make it possible to eradicate or, at least, control the potentially dangerous aspects of this ubiquitous communication (e.g., related to security, trust and the protection of personal data; reliability and dependability); to extend longevity, boost economy and improve the quality of life.

For example, in the context of the small cell deployments, emerging vulnerabilities are the following:

- Devices can be compromised
- Can be turned into IMSI catcher
- User data can be eavesdropped/ modified
- Denial of Service attacks
- Call fraud
- Privacy issues
- Considerations of security for use of technologies use across operators.

In another scenario, the technological push will naturally lead into the new wireless communication era. A possible barrier to fulfill the technological promises bringing about long delays into technological adoption can be expected from obstacles posed by negative public sentiment, legal barriers and lack of funding.

Such opposition may come from lack of education or understanding of how technologies work, as well as legitimate safety and ethical concerns.

To this end, GISFI has undertaken work to perform a study [11] on the telecom security policy addressed to Indian wireless network operators and to network equipment manufacturers/ vendors (both to Indian or Global organizations). This is very much in line with the recommendations of Department of Telecom (DoT), Government of India (GoI), of April 2013, stating that "...the operators shall integrate only those network elements into their telecom networks which have been tested and certified from (Government) authorized and certified agencies/labs in India."

The greatest obstacles entail moral or ethical considerations involving the manipulation of personal and national security data related to wireless communications for critical infrastructures. Significant developments and breakthroughs in new technologies have produced raging success in the wireless and ICT field, but many of the most novel and effective solutions (e.g., cognitive radio networks; spectrum and infrastructure sharing) are slow to appeal with consumers, government agencies, advocacy groups and in some cases, entire regions of the world.

2.3 Green ICT

The ICT industry is a contributor to the phenomenon of global warming, with a current footprint of 3% of the consumed energy. However the energy consumption of ICT is rising at a rate of 16-20% per year. It is also shown that 90% of the global cellular network power consumption belongs to the network side and, of that, 97% is consumed by the Radio Access Network [12]. Only recently has the idea of energy efficiency come to the attention of researchers, authorities and mobile operators. In the past, mobile networks had been designed to optimize spectral efficiency or capacity, but not energy efficiency. Good energy and spectral efficiency is translated into economics gain, because more traffic is sent over the network, traffic that generates revenue. Value added services like multimedia services and entertainment services increases the revenue that can be generated from the spectrum. Another efficient use of the spectrum is to "borrow" the unutilised spectrum a primary operator has to a secondary one as long as the primary user does not need it. This approach is another good revenue opportunity for primary operators of the spectrum.

In order to obtain any degree of energy saving, the attention should focus on the radio access side of the cellular network. Then the challenge is to design flexible techniques and algorithms in order to reduce the base station power consumption. Authors in [13] have identified two ways to reduce the energy consumption of Base Stations: (1) by having energy-aware components in BSs and (2) by energy-aware deployment strategies of BSs. However, the paper does not elaborate on those strategies.

Within GISFI, activities span energy efficiency issues related to greening of the telecom infrastructure as well as related to more energy-efficient utilization of the spectrum. These activities are undertaken within the corresponding Working Groups on Green ICT and Spectrum.

In India, TRAI [4] recommends that all telecom products, equipments and services in the telecom network should be energy and performance assessed and certified with a 'Green Passport' utilizing the Energy consumption Rating (ECR) and energy 'passport' determined by the year 2015. However, at this point of time telecom services have been excluded from the scope of the proposed 'Green Passport'. Telecom equipments are the current subject of focus, since adoption of energy-efficient telecom equipments would lead to overall energy efficiency of network and services [14]. Further, the ICT infrastructure used in telecommunication core networks also needs to be certified for energy-efficient operations. GISFI provides a generic definition of telecommunications equipment efficiency as the ratio of power to the useful work done[15]. Useful work and power are terms that should be defined in each supplemental standard. Additionally, in the GISFI supplemental standards, the following would be defined:

a. The scale of the energy efficiency metric for each class of equipment so that the metrics for similar class of equipment may be made comparable.
b. The lower the value of the energy efficiency metric, the more energy efficient is the metric compared to others of similar configuration.

GISFI has produced three technical reports (TRs) on Green ICT till date, the most relevant of which are the TR on Metrics and Measurement Methods for Energy Efficiency (GISFI TR GICT.105 [8]) and the draft technical specification (TS) on the Metrics and Measurement Methods for Energy Efficiency: General Requirements. GISFI's approach is not to reinvent the wheel, but re-use existing technical specifications from international standards wherever applicable and develop specifications if a clear technical gap is identified in the Indian context. As part of its efforts, GISFI has produced the following:

• Classification and prioritization of telecom equipments;
• Survey of International GICT standards and best practices;
• Recommended international standards with Indian requirements;
• Recommended GICT metrics and methods for measurement;
• Recommended assessment procedure and scales for rating.

2.4 Spectrum

In general, mobile services can use spectrum upto 6 GHz due to requirements of building penetration, multi-path, non line-of-sight, etc.

Hence, globally, spectrum managers are moving the existing line-of-sight, fixed links in frequency bands below 6 GHz, to higher frequency bands or alternate physical media like OFC, etc. In general, radio use in higher frequencies is a lot lighter, and currently, this gets a lot of attention as a potential way to realize 5G technologies.

In order to create a communication system that is revenue-rich, the 5G technologies must be realized so that they enable human-centric value-added services and applications. This will highly depend on cellular machine-to-machine (M2M) connectivity. Use of the millimeter wave communication links can add to the capacity enabled by small cells, leading to a new type of cellular architecture where base station connectivity to devices or other base stations can be based on use of millimetre wave links [16]. Still, the high absorption rate of the millimeter wave electromagnetic signal (60 GHz and beyond, like the e-band at the 70 GHz for industrial applications, but also higher bands such as 140 GHz and 220 GHz) poses great challenges for their utilization in non - line of sight (LOS) and mobile connections. On the other hand the high directionality attained in this band can be used to increase spatial multiplexing.

White spaces are being demanded and proposed to be used for low power un-licensed, Wi-Fi type usages, along with use of cognitive radio.

2.5 The Human-Centric Paradigm

In order to fulfil the needs of the modern user, 5G should offer human-centric connectivitywith data rates in the Tbps range and a coverage extending from a city region, to a country, the continents, and the world. The human-centric aspect will enable that connection traffic and application needs will grow to demand in turn 5G connectivity, which will facilitate deployment of 5G technologies and equipment.

Today, India is the second largest country in terms of mobile user base next to China[17]. Internationally, India is also considered the most potential market in mobile. It has been predicted that India will become the world's third largest smartphone market by 2017 after China and U.S [17] based on the 67 million smartphone subscribers at the time of the report, which is 6% of the total subscribers in India, growing at the rate of 52% YoY.

Within GISFI the human-centric paradigm and required connectivity are reflected within the work of the Working Groups on Internet of Things (IoT) and Cloud and Service Oriented Networks (CSoN).

The WG IoT has defined several use cases to help define the technical and usage requirements for human-centric connectivity and applications. 5G should seamlessly bridge the virtual and physical worlds offering the same level of all-senses, context-based, rich communication experience over fixed and wireless networks to users. As a first objective in reaching this vision, there should be no limits on wireless data rates. In a human-centric 5G scenario, each user should be provided with ubiquitous personalized network access at very high sustainable data rates approaching the Ethernet current state-of-the-art (10+ Gbps). A ubiquitous and pervasive wireless network offering a sustainable 10 Gbps bit rate (reaching up to 1 Tbps in a bursty mode) can be used as an alternative to Ethernet, and an access network to Tbps fiber networks.

Because 5G will be a plethora of interworking technologies governed by separate specifications, it is important to find technology solutions and standardize the interconnectivity for end to end telecom service provisioning across technologies and operators. Part of the activities within the WG on CSoN, have focused on this challenge [8].

The challenges related to the realization of the 5G end-to-end scenario are concerning the architecture design; the wired network segment transport protocol and the wireless/wireline functional interface nodes.

An optimized end-to-end architecture should be designed with the aim to provide excellent system performance in terms of bandwidth, utilization of network resources, reconfiguration capability, quality of service (QoS) and cost efficiency as well as survivability and security. Use of the cloud computing benefits and their integration is crucial in the context of the enormous volume of bursty traffic generated by the 5G solution.

In order for a cloud system to support the large flow of real-time data for human-centric 5G applications, the following general cloud requirements can be identified.

- Related to data rates and latency: The data should be pushed out to the cloud, and then pushed to the user because polling requires too much time and uses too much bandwidth. Further, the data needs to flow quickly and effortlessly through the system, through a real-time database, which means that it should be able to stay in its simplest format.
- Related to security: the 5G application should be able to stream its data into the cloud without exposing its data to the dangers of communicating it over the Internet. Further, support of different user types accessing a single service should also be enabled. Resource configurations should also be performed in a secure way for each service across multiple cloud infrastructures.
- Related to semantics: the information provided by different sensors or other 5G devices is usually in a specific format allowing it only to be used

by a proprietary application or to be controlled only by specific systems. But the provided information could be used by many other applications. It is therefore necessary for open interfaces and data formats to allow for using the different sensor outputs in applications, programming etc.

- Related to mobility: Specific and smart mechanisms should exist for controlling and exploiting the mobility of real-world entities and attached 5G devices.

3 A Vision of 5G

Networks will be the single, most indispensable element of future wireless connectivity, building an infrastructure of large-scale, complex and highly networked systems whose efficiency, sustainability and protection would require intelligent, interoperable and secure ICT solutions and novel business models. Users are not only humans, but also machines. At the core of 5G applications are machine-to-machine (M2M) communications enabled by technological breakthroughs in a plethora of scientific areas, wireless and wired communications, artificial intelligence, Internet, robotics, space, and so forth. This interworked 5G technological landscape is shown in Figure 3.

Always-on capacity, connectivity, and pervasiveness are key characteristics of the future 5G communication scenario that drive the emergence of new environments that evolve from the gradual development and combination of present day cellular communications and related advances (e.g., small cells), Internet of Things (IoT), M2M, and cloud computing, and towards a more advanced vision of fully reprogrammable mobile devices, able to communicate with each other autonomously based on a given event context and part of a scale-free self-organized communication systems.

There are three main aspects to realizing 5G communication systems.

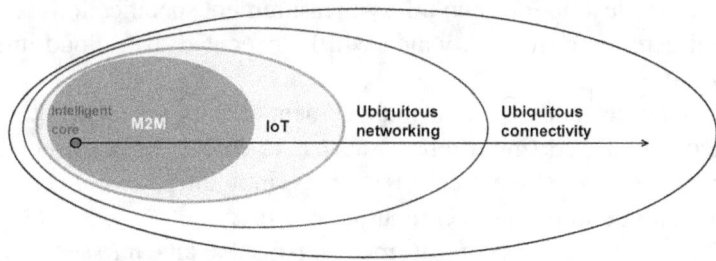

Figure 3 Plethora of technologies to deliver 5G services and applications.

- **Enabling the 5G intelligent core.** M2M and IoT are key for realizing the intelligent core, which in turn is key for enabling seamless ubiquitous networking and connectivity in a 5G context. In relation to the specifics of the Indian scenario, this intelligent core will be a turning point for bringing mobile services also to the Indian rural areas [18]. M2M and IoT are the key enabling technologies for a pervasive and always-connected 5G mobile services. Research challenges to fully deployable intelligent core are related but not limited to handling the big data collected through M2M and IoT communications (e.g., heterogeneous gateways, energy efficiency, decentralization of routing, naming and addressing), and to security, privacy and trust.
- **Enabling ubiquitous connectivity.** This feature has two aspects. On one hand, technical challenges relate to sufficient coverage range even in a scenario of very high mobility and data rates, and on the other, to moving application from device to device without any content interruption. Use of millimeter wave links as discussed previously, novel multiple antenna concepts, virtualization, small cell deployments, novel spectrum usage methods, are some of the key research enabling areas for ubiquitous connectivity.
- **Enabling ubiquitous networking.** This means that regardless of how many access networks are integrated for connectivity purposes, the quality of the delivered service must be retained end-to-end. In view of increased importance of the cloud computing concept for supporting the big data originating from the intelligent 5G core, end-to-end ubiquitous networking will require interoperable decentralized service-oriented mechanisms with support of real-time interactions. And end-to-end ubiquitous networking architecture will blend concepts and techniques from the three main 5G elements (see Figure 3), for example:

  o Service descriptions and utility measurement specifications for intelligent core services, which will be executed over cloud infrastructures.
  o Scheduling and resource management techniques that take into account shared (multi-tenant) access to (heterogeneous location-dependent) intelligent core resources, in addition to conventional (location-independent) computing resources.
  o Security and access controls mechanisms that take into account the sharing of resources (e.g., physical and virtual).

4 Conclusions

5G is a concept where personalization meets connectivity and networking technological innovations by integrating under one interoperable umbrella leading technologies, such as advanced M2M communication technologies, cooperative and cognitive radio and networking technologies, data mining and decision-making technologies, security and privacy protection technologies, cloud computing technologies, with advanced sensing and actuating technologies. 5G bundles multi-radio, multi-band air interfaces to support portability and nomadicity in ultra-high data-rate communications using novel concepts and cognitive technologies.

GISFI is currently undertaking 5G standardization by defining recommendations for the realization of the WISDOM concept.

References

[1] International Telecommunication Union (ITU), www.itu.int.
[2] Global ICT Standardisation Forum for India (GISFI), www.gisfi.org.
[3] Government of India, Department of Telecommunicaitons, "National Telecom Policy 2012," http://www.dot.gov.in/telecom-polices/national-telecom-policy-2012
[4] Telecom Regulatory Authority of India (TRAI), "'Highlights on Mobile Data Subscription in India as of January 2013," Press Release Report, New Delhi, March 2013, www.trai.gov.in.
[5] C.-I.Badoi, N. Prasad, V.Croitoru and R. Prasad, "5G Based on Cognitive Radio,"Springer International Journal on Wireless Personal Communications (2011) 57:441-464. DOI 10.1007/s 11277-010-0082-9.
[6] R. Prasad, "Future Networks and Technologies Supporting Innovative Communications," in Proceedings of IEEE IC-NIDC2012, Bejing, China, September 2012.
[7] R. Prasad, "High Impact Emerging Technologies for Beyond 2020 Wireless Communications," Proceedings of WWRF, October 2012, Berlin, Germany.
[8] Global ICT Standardisation Forum for India, GISFI, www.gisfi.org.
[9] 3GPP TR 36.814 V.9.0.0. (2010-03), "Further Advancements for E-UTRA physical layer aspects," Release 9.
[10] 3GPP TS 32.101 V11.1.0 (2012-12), "Telecommunication management; Principles and high level requirements", Release 11.
[11] WG Security, GISFI_SP_201303376, "(Telecom Security) Policy Study and Proposals," www.gisfi.org, March 2013.
[12] E.C. Strinati and L. Herault, "Holistic approach for future energy efficient cellular networks," Elektrotechnik & Informationstechnik, vol. 127, Nov. 2010, pp. 314–320.
[13] Arnold, O.; Richter, F.; Fettweis, G.; Blume, O., "Power consumption modelling of different base station types in heterogeneous cellular networks," Future Network and Mobile Summit, 2010 , vol., no., pp.1,8, 16-18 June 2010.

[14] WG Green ICT, GISFI_GICT_201301389 , "Approach towards Implementation of Green Telecom in India,"www.gisfi.org, January 2013.
[15] WG Green ICT,GISFI_GICT_201301357, "Metrics and Measurement Methods for Energy Efficiency: General Requirements," www.gisfi.org, December 2012.
[16] Millimeter-Wave Transceiver Technologies for 5G Cellular Networks – http://global.samsungtomorrow.com/?p=24093#sthash.x3KTbC9y.dpuf.
[17] Indian Mobile Landscape 2013, http://www.iamwire.com/2013/06/indian-mobile-landscape-2013/, June 2013.
[18] Confederation of Indian Industry-CII, "Machine-to-Machine: Vision 2020 Is India ready to seize a USD 4.5 trillion M2M opportunity?", TeleTech 2013 report, www.deloitte.com/in

Biography

Ramjee Prasad is currently the Director of the Center for TeleInfrastruktur (CTIF) at Aalborg University, Denmark and Professor, Wireless Information Multimedia Communication Chair.

Ramjee Prasad is the Founding Chairman of the Global ICT Standardisation Forum for India (GISFI: www.gisfi.org) established in 2009. GISFI has the purpose of increasing of the collaboration between European, Indian, Japanese, North-American and other worldwide standardization activities in the area of Information and Communication Technology (ICT) and related application areas. He was the Founding Chairman of the HERMES Partnership - a network of leading independent European research centres established in 1997, of which he is now the Honorary Chair.

He is the founding editor-in-chief of the Springer International Journal on Wireless Personal Communications. He is a member of the editorial board of other renowned international journals including those of River Publishers. Ramjee Prasad is a member of the Steering, Advisory, and Technical Program committees of many renowned annual international conferences including Wireless Personal Multimedia Communications Symposium (WPMC) and Wireless VITAE. He is a Fellow of the Institute of Electrical and Electronic Engineers (IEEE), USA, the Institution of Electronics and Telecommunications Engineers (IETE), India, the Institution of Engineering and Technology (IET), UK, and a member of the Netherlands Electronics and Radio Society (NERG), and the Danish Engineering Society (IDA). He is a Knight ("Ridder") of the Order of Dannebrog (2010), a distinguished award by the Queen of Denmark.

Spectrum Challenges for Modern Mobile Services

P. K. Garg, T. R. Dua and Ashok Chandra

Global ICT Standardisation Forum for India (GISFI)
www.gisfi.org

Received July 2013; Accepted August 2013

Abstract

Proliferation of wireless services and demand of greater mobility has exerted pressure on spectrum manager to make available additional spectrum in lower bands. With current spectrum management practice, no spectrum below 3 GHz is free/readily available for new services. Spectrum sharing is one solution to make available additional spectrum for new services. Sharing improves spectrum utilisation and provides enormous possibilities. This paper gives an idea about different aspects of spectrum sharing & spectrum trading and emphasized that Indian telecom market is good ground for trying various aspects of spectrum sharing. This paper is based on discussions held on utilisation of radio spectrum during several GISFI meetings held in India on standardisation.

Keywords: Radio spectrum, spectrum sharing, spectrum trading, cognitive radio, TV white space.

1 Introduction

The Radio Frequency (RF) Spectrum is a limited natural resource, which is governed by the laws of physics. Theoretically, the RF commences from approx. 9 KHz and extends upto 3000 GHz. However, in practical scenario, not every part of the spectrum is suitable for being used for all requirements.

Journal of ICT Standardization, Vol. 1, 137–158.
doi:10.13052/jicts2245-800X.12a2

For example, long distance communication in a single hop is possible through HF (high frequency) bands or through troposcatter systems only. Also, at present, technologies/ equipment are available for general use of spectrum upto approx. 85 GHz only. Mobile communication is feasible upto approx. 6 GHz band, with currently available technology. Even within this practically usable spectrum as per current available technologies, cost effective equipment for a particular application may be available for use in still smaller/ limited frequency bands. This is the practical limitation on the spectrum due to propagation and availability of suitable equipment.

Today, large portions of the radio spectrum stand assigned to the authorized users by governments over the last one hundred years. This spectrum allocation policy refers to command- and control mechanism in which the Government may decide the following: ration spectrum, specify technologies and services for spectrum use, put in strict mergers and acquisition (M&A) norms, and confer non-sharable rights to spectrum holders etc. [1]. This static spectrum allocation mechanism causes frequency bands to be insufficient in various times and locations. This seeming waste is commonly referred to as the problem of "regulatory overhead". Even if all such users are paying the spectrum charges, the opportunity cost and benefits for the society at large are not fully derived. Only some small parts of the radio spectrum are openly available to license-exempt users, and changing the status of a licensed radio spectrum to unlicensed can be crucial challenge. This process takes time and needs close cooperation between technological and technical bodies to achieve efficient deployment of the new policies.

Figure 1 [2] illustrates the radio spectrum and thebroad range of frequencies in wireless communication context. Most of the fractions of the radio spectrum are licensed to traditional radio communications systems. Beside, practical measurements, prove that most of the licensed bands either are unused or partially used at different geographical areas at most of the times According to the FCC's report the licensed spectrum band utilization range from 15% to 85% at different times and locations [3].

The regulation of the radio spectrum can be differentiated into four approaches explained in table below.

While the number of wireless connections and high data rate networks increase, spectrum demand and spectrum congestion become critical challenges in the forthcoming all-encompassing wireless world. In fact, throughput, high reliability, high quality of service, mobility, and diversity of wireless services, devices based on multiple wireless standards are more and more demanded. Hence, future wireless networks will face greater spectrum scarcity due to

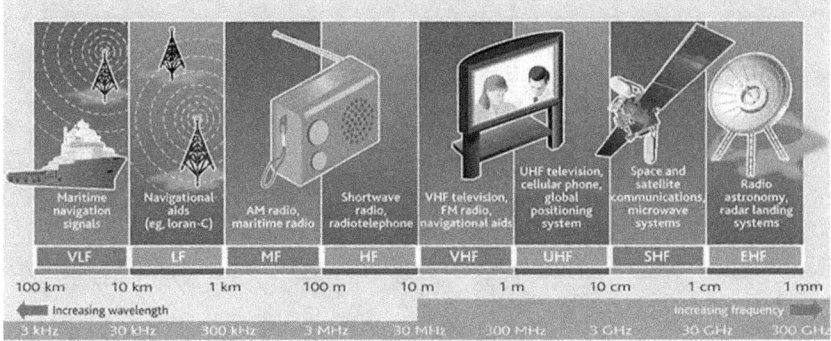

Figure 1   Radio spectrum [2].

the user's requirement such as high multimedia data rate transmission and diversity of communication technologies. The next section describes growing of spectrum demand in future wireless communication.

2 Increasing demands on RF Spectrum

Demands on RF spectrum are increasing extensively, for mobile services – broadband with ever increasing data rates (4G with 100 MBPS data rate) – futuristic 5G (ultra wide band – 1 GBPS)[4]. As per CISCO analysis of data usage [5], Mobile data traffic will grow at a CAGR of 66 percent from 2012 to 2017 mobile data traffic grew 70% in 2012 as compared to 2011, average data usage by smart phone devices is more than 50 times greater than the average data usage by basic feature phones, which comprise about 80% of the total connections/ devices globally. The predicted data usage in 2017 envisages average mobile data usage of about 7 GB per month by lap tops and netbooks, 2.5 GB by smartphone devices and more than 1 GB by non-smartphone devices. The overall mobile data traffic is expected to grow to 11.2 exabytes per month by 2017, a 13-fold increase over 2012. The 4G systems where deployed, have seen an average of 20 times more data traffic compared to existing systems. With cloud computing, the data usage is predicted to grow further. All this would need additional spectrum for meeting these growing requirements satisfactorily. Even now, need for additional spectrum of 500 MHz for mobile services is foreseen in near future in most of the countries.

As per Report ITU-R M.2078[6]: 'Estimated spectrum bandwidth requirements for the future development of IMT-2000 and IMT-Advanced', the predicted spectrum requirements for the mobile industry would be three times

Table 1 Regulation of radio spectrum (USA example).

Approach	Description	Application
Licensed Spectrum for Exclusive Usage	Protected through the regulator, transferable/flexible usage rights	Universal Mobile Telecommunication System (UMTS)
Licensed Spectrum for Shared Usage	Restricted to a specific technology	Digital Enhanced Cordless Telecommunication (DECT), public safety services, secondary usage
Unlicensed Spectrum	Available to all radio technologies working with specified standards, no right for protection from interference	Unlicensed National Information Infrastructure (U-NII) bands
Open Spectrum	Anyone can access any range of the spectrum, minimum set of rules define by standards	Low power underlay spectrum usage, Cognitive, frequency agile radios, cooperation-based, self-reconfiguring

Table 2 Future spectrum requirements.

Demand scenario	Total spectrum requirements (MHz)		
	2010	2015	2020
High Demand Setting	840	1300	1720
Low Demand Setting	760	1300	1280

Table 3 Identified/additional spectrum requirements.

Total (MHz)	Region-1(in MHz)		Region-2 (in MHz)		Region-3(in MHz)	
	Identified	Net additional	Identified	Net additional	Identified	Net additional
1280	693	587	723	557	749	531
1720	693	1027	723	997	749	971

the spectrum than in the last 20 years (Table 2). The report also gives details about already identified spectrum and additional requirement of spectrum for IMT and IMT Advanced services to meet this demand is different for all the 3 regions (Table 3).

At the same time, the requirements of RF spectrum for other services for public – both satellite based and terrestrial based communication services, broadcasting as well as various navigational services, are also increasing. Further, the requirements of RF spectrum from other captive users – strategic users as well as other government and private users are also increasing significantly.

2.1 Trends Contributing to Increased Demand for Mobile Broadband

Despite its short history, mobile broadband seems to have a higher growth impact relative to communication technologies, such as fixed and mobile telephony and the Internet [7]. There is growing evidence that mobile broadband has a considerable socio-economic impact for individuals, firms, and communities. Mobile broadband traffic and number of subscriptions are dramatically increasing since last few years. New types of mobile devices, such as smart phones, dongles and tablets and new user behaviours have emerged, as well as new applications have been created by users.

As the mobile broadband is providing benefits for society and economics, it is expected that data traffic increase and convergence between mobile and other services, such as e-health, e-education, will provide further benefits. The demands for multimedia (uses multiple forms of information content and information processing (e.g. text, audio, graphics, animation, video, interactivity), along with e-education, e-health, mobile commerce, mobile broadcasting/multi-casting, are some emerging telecommunication services to mobile users that can be provided by 'IMT' and 'IMT Advanced'.

The new generation of mobile broadband networks will support higher data throughput rates, lower latencies and more consistent network performance through a cell site [8]. This will increase the number of applications and devices that can benefit from mobile broadband connectivity, generating a corresponding increase in demand for mobile broadband from consumers, businesses. Some 'trends' contributing to increased demand for mobile broadband are:

1. New type of devices, such as smart phones, dongles, tablets
2. Mobile Internet usage is increasing
3. Huge increase of mobile applications
4. Video traffic is growing dramatically
5. Media rich social networks go mobile
6. Machine-to-Machine traffic is growing rapidly
7. More capable network – user experience improvement
8. Cost reduction and price decrease
9. Several policy initiatives to promote mobile broadband
10. Potential area to increase data traffic

2.2 Developments in the Field of Computers and Wireless Communications

The developments in the field of computer software and mobile communication facilities have complimented each other. The modern communication systems make very extensive use of computer hardware and software facilities. The enhanced computing powers and larger memory capacities have helped in development of better communication systems. At the same time, more efficient communication systems allow exchange of much larger amount of data, thus allowing greater exchange of software development capabilities between any parts of the globe. The resultant volumes and economies of scale have helped bring down the cost of equipment, allowing much faster expansion of telecom facilities across the globe. All these developments have exponentially increased the volume of data travelling across various networks, including wireless networks. This has necessitated more efficient use of all available and usable frequency spectrum.

3 Technological Developments in Spectrum Utilization

The evolution of technologies has gradually allowed use of higher frequency bands, thus increasing the availability of spectrum. For example, about 25 years ago, mobile communications were feasible in frequency bands upto about 400 MHz only. As compared to that mobile communications systems are now available upto approx. 6 GHz – increase in spectrum availability by almost 15 times for mobile services. However, most of the frequency bands upto 6 GHz are heavily in use for variety of other services. So the spectrum shortage continues.

Similarly, the line-of-sight systems, which were generally available/ usable upto 15 GHz about 25 years ago, are now available upto almost 100 GHz band. Hence, the spectrum availability for such systems has also increased significantly. However, the demand/ requirements have grown at a much faster pace, thus resulting in continued shortage of spectrum.

On another front, the miniaturization of electronic components & systems has helped more efficient usage of spectrum. In this connection, introduction of multiple frequency bands in single equipment, due to miniaturization, allows enhanced dynamic sharing of frequency bands and the efficiency of spectrum usage is largely increased.

Hence, it has been practically observed that increasing demands on spectrum have outstripped the practical availability of spectrum, despite higher frequency bands becoming available for use through technological developments. In reality, the demands on RF spectrum have always been more than its practical availability – hence the need for its regulation/ management in an effective manner so as to derive maximum benefit for the nation and humanity at large, in the international scenario. Optimum spectrum sharing results in greater efficiency of spectrum usage. Greater sharing of frequencies and bands allows more data to be sent by different users in the same amount of available spectrum.

4 Spectrum Sharing

The concept of spectrum sharing [9] is sharing of a frequency band by two or more than two radio communication services or applications. At international level, different bands of RF spectrum are shared among multiple services. Each country can choose one or more of such services for utilization within their country – subject to the condition that services of other countries do not get any harmful interference [10]. Protection criteria for sharing among different services and applications are decided at international level taking into account various technical factors. With the development of technology, these sharing and protection criteria also undergo change periodically. Sharing has basically three dimensions; frequency, time and location. Spectrum sharing is not a new phenomenon. It has been in practice since longback. The frequency reuse concepts in existing telecom network, operation of point to point link on same frequency at different locations, FDMA & TDMA etc are example of spectrum sharing. Sharing of spectrum among homogenous/ similar services & applications is relatively easier and leads to greater spectrum efficiency as compared to sharing among heterogeneous services and applications (sharing among largely different services & applications) is more complex as well as provides relatively lesser increase in spectrum efficiency. In general, the digitally modulated carriers can work satisfactorily with lower protection criteria.

A lot of work is carried out in various study groups of ITU-R [11] for establishing sharing criteria among different services and applications. These criteria play a crucial role in appropriate sharing of spectrum and these are periodically reviewed in the light of various technological developments. With the development of technology, the sharing criteria undergo improvement continuously, resulting in denser utilization of and more traffic/data being derived from the same spectrum. As large number of countries participates in ITU-R studies, the criteria adopted by study groups are relatively conservative. For usage within the country or during bilateral discussions between two countries, it is possible to accept more liberal criteria, allowing still better utilization of spectrum.

4.1 Spectrum Sharing – US Scenario

The idea of spectrum sharing in the United States really started gaining traction after the National Telecommunications and Information Administration(NTIA) issued a report in November 2010 that found 115 MHz of spectrum currently in the hands of the federal government that could be used for wireless broadband [12]. That kicked off lots of conversations over whether and how wireless carriers could share spectrum with the federal government. Another NTIA report, in March 2012, found that 95 MHz of spectrum currently in federal hands, the 1755–1850 MHz band, could be repurposed for commercial wire-less use, and as part of its review, the NTIA recommended both relocating federal users and sharing spectrum between federal agencies and commercial users [13]. In August 2012 the FCC granted permission to T-Mobile USA to test the concept of sharing spectrum between federal and commercial users in the 1755–1780 MHz band. AT&T Mobility, Verizon Wireless and T-Mobile USA recently inked an agreement with the Department of Defense to explore the possibility of sharing 95 MHz of spectrum that is currently used by the Pentagon and other federal agencies located in the 1755–1850 MHz band [14].Of late, AT&T and Verizon are on a buying spree, as they vie with each other in expanding their 4G LTE (long term evolution) networks. For example, in 2012, AT&T obtained approval to purchase 700 MHz and 2300 MHz band spectrum from the likes of NextWave Wireless, Comcast, Horizon Wi-Com and San Diego Gas & Electric Company. Verizon recently agreed to pay $3.6 billion to buy spectrum from a consortium of cable companies to augment its spectrum capacity. Sprint acquired Clearwire recently for its 2.6 GHz 4G spectrum [1].

The USA President's council of advisors on science and technology in their report last year, recommended about 1,000 MHz of federal spectrum holding to be released for shared access [15]. In a first ever move, the US military will experiment with sharing of spectrum that they use for aviation radars in the 3550–3650 MHz band for 4G LTE based indoor networks in hospitals [1].

The NTIA, in coordination with the Federal Communications Commission (FCC) and the Federal agencies, established a Spectrum Sharing Innovation Test-Bed (Test-Bed) pilot program to examine the feasibility of increased sharing between Federal and non-Federal users. This pilot program is an opportunity for the Federal agencies to work cooperatively with industry, researchers, and academia to objectively examine new technologies that can improve management of the nation's airwaves. The Test-Bed Pilot Program will evaluate the ability of Dynamic Spectrum Access (DSA) devices employing spectrum sensing and/or geo-location techniques to share spectrum with land mobile radio systems operating in the 410–420 MHz Federal band and 470–512 MHz non-Federal band [16].

4.2 Dynamic Sharing of Spectrum Using SDR and CR

Earlier the transmitter equipments were tuned to specific frequencies and provision for multiple frequencies/ channels/ carriers meant large extra cost. The development of 'Software Defined Radio (SDR)' allows the user(s) to use/ switch to different frequencies in a dynamic manner without large increase in cost of equipment;

Still the existing users were not comfortable with dynamic sharing and anticipated harmful interference to their systems, when they would need to use a particular frequency/ carrier / channel. The development of 'Cognitive Radio (CR)" along with Software Defined Radio (SDR) [17], has taken care of this concern of existing users to a large extent. In case of Cognitive Radio (CR), the system would first listen and if the channel is occupied, it would switch to some other vacant channel and use it. Also, such systems would regularly (& frequently) sense the carrier for other usages and the moment it finds some other licensed usage, it would switch to some other vacant channel/ carrier, thus causing no harmful interference to existing/ earlier licensed user(s).

Hence, dynamic sharing of spectrum is being adopted increasingly. It is a combination of administrative (regulatory), technical and market based techniques to enhance the efficiency of spectrum utilization [18]. Such dynamic sharing leads to enhanced and optimal usage of the RF spectrum.

Many strategic users as well as other public safety users (e.g. Disaster Management authority) need a small amount of spectrum for their regular use. However, in case of emergency/ public safety situation, they need much larger amount of spectrum to take care of the situation. It is not desirable to block their total spectrum requirements for all times and throughout the country, as it would let large amount of spectrum lying unused as well denying other users the opportunity to derive societal benefit from such unused part of spectrum.

Hence, Spectrum regulators in more and more countries are encouraging dynamic spectrum sharing with public safety requirements. In USA,one block of spectrum in 700 MHz band has been auctioned with the condition that public safety services would be able to pre-empt its usage in case of any emergency/ public safety requirements [19].

4.3 Dynamic Sharing of White Spaces

Another area of dynamic spectrum sharing, which has been of large interest recently, is commonly known as 'White Spaces' (in TV band).The TV broadcasters normally plan to repeat the same channel/ carrier at relatively larger distances, to avoid any interference especially at the edges/ border of the coverage areas of two adjacent broadcast transmissions on the same channel. However, there are large areas near the edge of the theoretical coverage area of any particular TV transmission, where there are very few people receiving the program. Thus the spectrum for that channel is not used effectively in such areas.

The broadcasters in general are quite sensitive, even touchy, for protecting the reception of their signals even beyond theoretical coverage areas. Hence, only low power systems, that too on pre-emptive basis can be considered for shared usage with the TV spectrum. With the passage of time and gaining of adequate confidence among all users, including broadcasters, gradually higher power levels for other sharing systems would be feasible.

5 India Scenario

Spectrum sharing to some extent has been allowed in India in controlled manner with regulator's approval in non-commercial bands. However, spectrum sharing in commercial bands has not yet been opened. But the Government committed in National Telecom Policy 2012[20] document that *"To move at the earliest towards liberalisation of spectrum to enable use of spectrum in any band to provide any service in any technology as well as to permit spectrum pooling, sharing and later, trading to enable optimal utilisation of spectrum through appropriate regulatory framework."*

The telecom industry in the country has witnessed a phenomenal growth in the last decade, mainly for mobile telephones. With 900 million mobiles phone connections at the end of March, 2013, India is today the second largest and fastest growing telecom market in the world in terms of number of wireless connections. It continues to grow at an average rate of 7 to 8 million connections a month. A significant part of this growth is now taking place in smaller cities and rural areas. Cellular Operator Association of India (COAI) has projected subscriber growth to 1516.8 million in 2020 with almost 45% increase from 2012 subscribers figure (Table 4).

India is unique in its excessive fragmentation of spectrum holdings, and hence requires more innovative approaches in spectrum management. The average spectrum holding by each telecom operator is 10 MHz across all bands (i.e. 800, 900, 1800, 2100 MHz paired bands), about one-fourth of the international average [1]. To some extent, spectrum sharing has been introduced by allowing 2G intra-circle roaming and the liberalisation of spectrum use in recently conducted auction. But these steps are not sufficient to accelerate the market. Operators need more spectrum commensurate with their counterparts in other countries. But due to the Government's inability to allocate more

Table 4 Wireless subscriber projection.

	2011	2012	2013	2014	2015	2020
Population (in Million)	1218	1233	1249	1265	1281	
Subscribers (in Million)	923.8	1049.1	1134.5	1185.3	1217.1	1516.8
Teledensity	75.85	85.09	90.83	93.70	95.01	

spectrum, it is better that Government allows operators to manage spectrum through spectrum sharing/trading.

It is worth to note that not a single operator got 3G spectrum in all the 22 service areas. Operators decided to share spectrum (without transfer of spectrum rights) through mutual agreements without Government permission, thus creating an "unofficial" secondary market. This is an excellent example of operator enterprise. However, the Government has just stopped this practice. It is time the Government legitimises it *ex-post* by allowing spectrum sharing and trading for the benefit of all which consequently provide better quality of service for subscribers, optimal utilisation of spectrum and increase in Government revenue [1].

5.1 Initial Investigation of Frequencies Bands for Flexibility

The following bands are being investigatedin many countries for more flexibility regarding additional spectrum. Total bandwidth under consideration is about 1330 MHz.

- 470–862 MHz: presently used for broadcasting services;
- 880–915 MHz / 925–960 MHz as well as 1710–1785 MHz / 1805–1880 MHz: these bands form the 900/1800 network for GSM mobile services;
- 1920–1980 MHz/ 2110–2170 MHz: these bands are used for third generation (3G) mobile services;
- 2500–2690 MHz (the 2.6 GHz band): this band is used for Broadband Wireless Access (BWA) services;
- 3.4–3.8 GHz: this band is proposed for use by IMT and IMT Advanced services.

5.2 Policies on Spectrum Utilization

Light Licensing
In this utilisation technique all stations must be registered in defined database spectrum map. Light Licensing is a novel concept in the US in the 3,650 to 3,700 MHz range(and in 70/80 GHz bands), which allows systems to share spectrum on a co-primary basis, whereby geographic location and signal sensing mechanisms are employed to ensure that no interference is caused among the involved systems (Fixed, FSS or Mobile). The IEEE 802.11y Working Group is one standard body that has focused on such concepts [21].

Spectrum fragmentation

Spectrum fragmentation is a further reason that new techniques are needed to exploit spectrum gaps and white spaces among the fragmented spectrum assignments [22]. The approach is based on a joint simultaneous use of different frequency bands in the frequency range of about 400 MHz up to about 6 GHz including the unlicensed frequency bands for WLAN applications.

Spectrum Harmonisation

Spectrum Harmonisation means defining technical conditions, including spectrum, band plan and technology, at a global and regional level, to ensure efficient spectrum use, seamless services over wide areas including roaming, system co-existence and global circulation of user equipment across borders.

Spectrum Liberalisation

Essentially, liberalisation of spectrum means the removal of technology restrictions to enable new access technologies to be deployed within the same band or bands as existing and legacy technologies – for example, UMTS or HSPA could be deployed in spectrum bands where traditionally GSM, CDMA or TDMA has been used.However, such usages should not cause harmful interference to any other adjoining usage in the band or nearby frequency bands.

6 Spectrum Trading

Spectrum trading is another aspect associated with spectrum sharing. In fact, spectrum trading is the case of spectrum sharing, where commercial aspects are also involved. Spectrum trading permits the purchaser to change the use to which the spectrum was initially put while maintaining the right to use. In such cases, it may involve transfer of rights to use the spectrum, rather than transfer of the licence itself [23]. Spectrum trading generally leads to an economically more efficient use of spectrum. This is because a trade will only take place if the spectrum is worth more to the new user compared to the old user. Trading, viewed by many as the key step to be taken in the reform of spectrum management regulatory practice, is capable of unlocking the potential of new technologies and eliminating artificial scarcities of spectrum, which find expression in inflated prices for spectrum-using services. USA and many European countries have introduced/ allowed spectrum trading in certain specific bands only, which are in demand for commercial use, with specified conditions.

6.1 State of Spectrum Trading

Following the initial assignment of spectrum rights and obligations to users, whether by auction or other means, circumstances may change causing initial license holders to want to trade their rights and obligations with others. Today this is not possible in many countries. However, in a few countries such as the UK, secondary trading – the trading of spectrum rights after the primary assignment – is possible [24]. The possibility to trade radio spectrum is argued by many commentators to be a critical factor in the promotion of more efficient radio spectrum use. Furthermore, it is increasingly recognized that the flexibility afforded by trading is helpful for innovation and competitiveness.

The successful implementation of spectrum trading requires a commitment to change current view of regulatory bodies with a solid base in understanding new technologies and operating systems. Spectrum policies must address the incentives for innovation in order to promote spectrum's assignment flexibility while clearly establish the usage rights and obligations of those who use the spectrum to transmit or receive information. Furthermore, the spectrum flexibility also demands new approaches and practical methods for the monitoring compliance, enforcement and conflict resolution.

In Europe, the European Commission (EC) is taking the lead in promoting harmonized trading for radio spectrum where its use has a European dimension. Emphasis is being placed on certain bands below 3 GHz, where it is estimated that the net benefits from trade may be substantial. Despite fairly widespread recognition that the current regime of spectrum management operating in most of the European Union is not sufficiently flexible to achieve the Union's objectives in promoting competitiveness and innovation, thus far the pace of reform is slow, although some necessary steps have been put in place, and the European Commission is promoting liberalization across the EU [25].

Secondary spectrum market is already running in several countries like USA, UK, Australia, New Zealand and most of European Union countries. Though secondary spectrum markets have not been successful in Australia, New Zealand and some European countries due to "market thinness", or lack of sufficient participation, the activity level is very high in the US. Mobile operators have bought spectrum from each other as well as from broadcasters and other niche spectrum holders.

6.2 Spectrum Transaction methods

Spectrum transfer, proceeds by the transfer of licence rights and obligations and necessarily involves the grant of a new licence to the transferee. Transfer may be for all or part of the licence duration (short time or long time).

Spectrum leasing, proceeds by a Contract between the parties without the need for National Regulatory Authority(NRA)to issue a new licence. The incoming user (the leaseholder) is not issued with a licence by NRA but is authorised to use the spectrum for the period of the lease on the basis of a contract with a licensee (the lessor).

6.3 Duration of Usage Rights

Long Period: In the case of auctioned spectrum, or spectrum granted by means of comparative or negotiation procedures, the duration is longer, between 5 and 20 years. In Denmark the default duration for the granted usage rights is 15 years.

Short Period: In the case of frequency assignments with no selection procedure, they are generally granted for a short period, one to three years, but the licence is automatically renewed, if the license fee has been paid (e.g. Cyprus, Estonia).

6.4 Spectrum Trading Procedure [25]

The procedures can be broken up in several steps, essentially what happens before the transaction and what happens immediately after the transaction.

1. Notification of the intention to trade, spectrum holder might notify NRA in selling spectrum band.
2. Publication of information prior to the transaction, at this stage information of both parties and information on frequency and technical details need to be published.
3. Approval of transaction by NRA, publication of information on the effective transaction, information of the transaction must be made public, such as Geographic area, frequency, expiration, and so on.
4. Monitoring system, in most cases NRA has responsibility on monitoring transaction of spectrum usage rights.

6.5 Spectrum Trading Benefits

The main benefits of spectrum trading encompass:

- More efficient use of spectrum*
- More flexibility in spectrum management, including removal of rigidities in primary assignment.
- Ability to evaluate spectrum licences, and gain knowledge of market value of spectrum
- Facilitating market entry.
- Encouragement of innovation, enabling new technologies and market development.
- Speedier process, with better and faster decision-making by those with information.
- Increase in competition and reduced barriers to market entry.
- Reduction in administrative workload.
- Reassignment of spectrum from low economic value uses to high economic value uses.
- Allows efficient companies to expand and displace less efficient companies.
- Increasing opportunities for entrepreneurs to access spectrum to introduce innovative technologies and services.
- Reduction in the transactions costs of acquiring rights to use spectrum.
- Permitting more rapid redeployment and faster spectrum access for innovators and new players without the need for regulators to re-plan and re-farm spectrum.
- Allowing new technologies to gain access to spectrum more quickly.
- Opportunity (for existing operators) to sell unused or under-used spectrum and make more flexible use of spectrum.

6.6 Significant Concerns in Spectrum trading

There are several significant concerns in spectrum trading and liberalisation includes:

- Low spectrum trading activity
- Inefficient use of spectrum*
- High transactions costs
- Risk of increased interference
- Impact of spectrum trading on anti-competitive conduct
- Impact on investment and innovation

- Impact on international co-ordination / harmonisation
- Windfall gains
- Disruptive effect on consumers
- Reduced ability to achieve public interest objective

(*The spectrum usage efficiency varies with different operators. The spectrum trading generally would see the spectrum moving from an operator who has relatively lesser need (hence lesser spectrum efficiency), to another operator who is starved of spectrum due to larger amount of network traffic. Hence, such movement of spectrum through trading would result in more efficient use of relevant spectrum. However, all spectrum trading deals might not be undertaken for spectrum efficiency only. In some cases, an operator may buy part of the spectrum for his immediate needs and some more for his foreseen future needs (or total spectrum for his foreseen needs), when such spectrum is available and the buyer sees a business case for the additional spectrum. Such a deal might result in sub-optimum spectrum efficiency for some time, but eventually the buyer would plan his network to derive the maximum traffic and efficiency from his spectrum holding.)

7 Conclusions

Thus it can be seen that enhanced spectrum sharing results in greater efficiency of spectrum usage, which is essential – almost unavoidable necessity for this limited and scarce resource. Although, there is always some cost associated with it, yet the overall benefits to the society are much larger. Spectrum trading is also a kind of spectrum sharing, though it is not much successful in many countries but it would improve spectrum efficiency in both terms; technical as well as economic, if implemented in an objective manner, seeing the telecom scenario in India.

References

[1] V. Sridhar, Rohit Prasad. Nothing wrong with spectrum sharing. http://www.the hindubusinessline.com/opinion/nothing-wrong-with-spectrum-sharing/article 4992619.ece

[2] Radio Spectrum. http://kids.britannica.com/comptons/art-164539/Commercially-exploited-bands-of-the-radio-frequency-spectrum

[3] F. Akyildiz, W.-Y. Lee, M. C. Vuran, and S. Mohanty, "NeXt generation/dynamic spectrum access/cognitive radio wireless networks: A survey," *Computer Networks Journal.*

[4] Cornelia-IonelaBadoi, Neeli Prasad, Victor Croitoru, and Ramjee Prasad. 2011. 5G Based on Cognitive Radio. *Wirel. Pers. Commun.* 57, 3 (April 2011), 441–464. DOI=10.1007/s11277-010-0082-9.

[5] Cisco Visual Networking Index: Forecast and Methodology, 2012–2017.

[6] ITU-R M.2078Report. http://www.itu.int/dms_pub/itu-r/opb/rep/R-REP-M.2078-2006-PDF-E.pdf

[7] DimitriZenghelis. The Economics of Network-Powered Growth. www.cisco.com/web/about/ac79/docs /Economics_NPG_FINALFINAL.pdf

[8] Heterogeneous Network. http://www.ericsson.com/res/docs/whitepapers/WP-Hetero geneous-Networks.pdf

[9] ITU_R M Report. www.itu.int/dms_pub/itu-r/opb/rep/R-REP-M.2243-2011-PDF-E.pdf

[10] National Frequency Allocation Plan (NFAP) 2011 of India. www.wpc.dot.gov.in

[11] Study groups of ITU-R. http://www.itu.int/en/ITU-R/study-groups/Pages/default.aspx

[12] www.ntia.doc.gov/files/ntia/publications/ntia_fcc_letter_115_mhz_01192011.pdf

[13] http://www.ntia.doc.gov/files/ntia/publications/ntia_1755_1850_mhz_report_march2012.pdf

[14] http://www.fiercewireless.com/story/att-verizon-t-mobile-forge-pact-explore-spectrum-sharing-government/2013-01-31

[15] http://www.whitehouse.gov/sites/default/files/microsites/ostp/pcast_spectrum_report_final_july_20_2012.pdf

[16] http://www.ntia.doc.gov/category/spectrum-sharing?page=2

[17] Simon Haykin. Cognitive Radio: Brain-Empowered Wireless Communications. IEEE Journal on Selected Areas in Communications, Vol. 23, No. 2, February 2005

[18] ICT regulation tool kit on spectrum sharing.www.ictregulationtoolkit.org/sectionexport/word/5.4

[19] The First Responder Network and Next-Generation Communications for Public Safety: Issues for Congress http://www.acuta.org/wcm/acuta/legreg/061213a.pdf

[20] National Telecom Policy 2012 of India. www.dot.gov.in

[21] IEEE 802.11y. http://www.wi-fi.org/sites/default/files/WFA_11y_Primer_final.pdf

[22] Visions and research directions for the Wireless World. http://www.wwrf.ch/files/wwrf/content/files/publications/outlook/Outlook5.pdf

[23] Radio Spectrum Management for a Converging World. http://www.itu.int/osg/spu/ni/spectrum/RSM-BG.pdf

[24] Ofcom spectrum trading guide. http://stakeholders.ofcom.org.uk/binaries/spectrum/spec trum-policy-area/spectrum-trading/tradingguide.pdf

[25] European practices in trading of spectrum usage rights. www.cept.org/.../**European** %20practices%20in%20**trading**%20of%20S

Biographies

Mr. P. K. Garg is presently a Member of the Radio Regulations Board (RRB) of International Telecommunication Union (ITU), Geneva and former Wireless Adviser to the Government of India (head of national spectrum management organization of India). In 2010, he was internationally elected for the second term on the RRB. He has been elected as Chairman of the Board for the year 2013.

The Board resolves various issues concerning implementation of ITU Radio Regulations, which is an international treaty among member countries of ITU for planning and use of the RF spectrum, as well as resolution of disputes, if any.

An engineering graduate with more than 40 years of experience in radio-communications and spectrum regulations – nationally and internationally, Mr.Garg has widely travelled since 1980, for participation in various international and regional conferences of ITU, APT (Asia Pacific Telecomunity) and other international & regional bodies connected with telecommunications. He has also represented India on the Governing Council of ITU and served as the Senior UN & ITU Expert in Radio Frequency Spectrum Management in Saudi Arabia.

He has chaired as well as spoken at large number of national and international conferences and seminars in the field of wireless communications and RF spectrum management, both in India & abroad. He has published many papers on wireless communication and spectrum related issues.

He presently holds the following positions at international and national level:

1. Chairman, Radio Regulations Board, ITU, Geneva
2. Chairman, GISFI Spectrum Group
3. Fellow of Institution of Electronics & Telecom Engineers (IETE) India
4. Chairman, CII Sub-Committee on Telecom Licensing and Spectrum
5. Vice President, ITU APT Foundation of India
6. Executive Member, PTC India Foundation
7. Member, Telecom committee of ASSOCHAM India
8. Founder Member, NGN forum of India
9. Eminent Engineer for 2003 by Institution of Engineers (India), Delhi

T. R. Dua An engineering graduate with diploma in Business Management and export Marketing. Have an experience of over 35 years in the telecom sector. Experience includes all facets of telecom be it Product Development, Business Development, Telecom Licensing, Regulatory issues with respect to interconnection / roaming / unified licensing and infrastructure sharing / Mobile Number Portability, Spectrum Management, Spectrum Related Issues Like Spectrum Pricing, Efficient Utilization and Spectrum refarming etc. Have also been involved very closely with number of Joint Ventures / Technical Collaborations.

All these years he has held very prestigious positions as Director in leading telecom companies like Bharti Airtel Ltd., Shyam Telecom Ltd., Cellular Operators Association of India, Vice Chairman, Global ICT Standardisation Forum for India and Senior Director TAIPA. Besides other consulting assignments.

Have many first to credit like finalization of joint ventures, technical collaboration, introduction of new product, launch of cellular services in India and finalization of licence agreements / interconnect agreements.

Work very closely with institutions like ITU/APT/WWRF with regards to the spectrum matters. Very actively involved with various professional institutions / associations to promote the interest of telecom sector and published many papers in telecom.

Presently holding the following position in various Professional Institutions:

1. Vice Chairman – GISFI
2. Co Chairman – ITU APT Foundation of India
3. Vice President – PTC India Foundation
4. Fellow of Institution of Engineers (India)
5. Fellow of Institution of Electronics & Telecom Engineers
6. Member – Computer Society of India
7. Member – Indian Science Congress
8. Member – Optical Society of India
9. Member – WWRF
10. Member – FICCI/ASSOCHAM/CII Telecom committee
11. Chairman CII committee on GREEN ENERGY
12. Member - Amity Telecom Vision Board
13. Member Core Group "Mobility For Life " EU Project

 **Dr. Ashok Chandra** An Indian Engineering Services officer of 1976 batch did his Post graduate in Electronics and PhD in Electronics and Doctorate of Science (D.Sc.) in Radio Mobile Communications. He has worked as a Guest Scientist on DAAD Fellowship at the Institute of High Frequency Technology, Technical University (RWTH), Aachen, Germany during 1995 and 1999. During 2002, worked as a Guest Scientist on DAAD Fellowship at Bremen University, Bremen (Germany), where he undertook a series of research studies in the area of radio mobile communications. Dr. Ashok Chandra is having Technical Experience of about 35 years in the field of Radio Communications/Radio Spectrum Management including about 7 years of experience dealing with Technical Education matters of Indian Institutes of Technology, Indian Institute of Science etc. particularly their various research projects also in the areas of telematics/radio communications and presented over 25 research papers at various International Conferences in the areas of EMI, Radio Propagation etc. He has visited various technical Institutions and Universities namely Technical University of Aachen, Germany; Aalborg University, Denmark; Bremen University, Germany; and University of Lisbon, Portugal etc. and took several lectures in the area of radio mobile communications at Bremen University and Aalborg University. He has chaired various Technical Sessions at the International Conferences on Wireless Communications.

Dr. Chandra recently superannuated from the post of Wireless Adviser to the Government of India. In his responsibility as Wireless Adviser, he was associated with spectrum management activities, including in spectrum planning and engineering, frequency assignment, frequency coordination, spectrum monitoring, policy regarding regulatory affairs for new technologies and related research & development activities, etc. He was also associated with implementation of a very prestigious World Bank Project on National Radio Spectrum Management and Monitoring System (NRSMMS). This project includes automation of Spectrum Management processes and design, supply, installation/ commissioning of HF/VHF/UHF fixed monitoring stations; V/UHF mobile monitoring stations; and LAN/WAN communications network etc. He served as a Vice-Chairman, Study Group 5 of International Telecommunications Union (ITU)-Radio Sector. He has represented India to a large number of ITU meetings including World Radio Conferences (WRC). He served as Councillor from Indian Administration in ITU Council. Currently

Dr. Chandra is Adjunct Professor at the Indian Institute of Technology (IIT), Bombay.

His area of interest includes Technical Education, Radio Regulatory affairs for new technologies, Radio mobile communications, dynamic spectrum management and Cognitive Radio.

Spectrum Trading in India and 5G

Purnendu S. M. Tripathi and Ramjee Prasad

Center for TeleInFrastruktur (CTiF), Aalborg University, Denmark;
email:{in_psmt, Prasad}@es.aau.dk

Received July 2013; Accepted August 2013

Abstract

Currently radio spectrum is largely managed through Command and Control method. Public mobile services require spectrum below 3 GHz for providing cost effective services. The existing method has created artificial shortage of spectrum especially below 3 GHz. Spectrum trading is a new concept in which service providers are permitted to purchase spectrum from the market to fulfil their requirements. Spectrum trading has not yet been permitted in India. This paper provides an overview of possibilities of spectrum trading in India and concludes that necessary ingredients are present in India for spectrum trading and it could provide a boost to the Indian telecom sector. Further, it will also discuss spectrum issue related with 5G in the direction of millimeter waves.

Keywords: Radio spectrum, 2G/3G services, spectrum trading, NFAP 2011.

1 Introduction

The journey of mobile communications started in the early 1980s with first generation (1G) mobile communications with analog technology primarily for voice only. Now we are in fourth generation mobile communications with all IP based packet switched network, seamless mobility and high data rate with 1 Gbps peak data rate for downlink (DL) and 500 Mbps for uplink (UL) [1]. The spectrum requirements upto 4th generation mobile communications can easily be met upto 6 GHz. Although, the bands for IMT services have been

Journal of ICT Standardization, Vol. 1, 159–174.
doi:10.13052/jicts2245-800X.12a3
© 2013 *River Publishers. All rights reserved.*

identified by the ITU but these bands are not free. It has already been assigned long back for other usage. The current spectrum management process does not allow spectrum holders the flexibility to respond quickly to changes in the market demand and technology, resulting spectrum lying underutilised, which creates artificial scarcity [2].

The 2G in 900/1800 MHz band and 3G mobile services in 2.1 GHz band are running successfully in India [3]. 4G services in 2.3-2.4 GHz band has also entered into Indian market but has not yet rolled out satisfactorily due to limited coverage. In the USA and European countries, 4G services are running successfully in lower bands by allowing use any technology in any band[1]. The assigned spectrum is contiguous, in case of shortage, operators can get additional spectrum from the market through spectrum trading. However, the situation in India is just reversed, here assigned spectrum is un-liberalized (except acquired through auction) and highly fragmented. Therefore, a flexible environment is required to be created for further expansion of telecom sector so that 4G services can be reached at every corner of India.

Spectrum trading [4] is an option through which flexibility can be increased and spectrum, assigned to a particular service, can be easily transferred for other usage. Spectrum trading improves the efficiency and facilitates new services to enter in the market. Spectrum trading has already been implemented in most of the countries but it has not yet been introduced in India. Presently, spectrum has been delinked with license and available through market mechanism only. At this stage, Indian market seems to be mature enough for spectrum trading.

The paper is organized as follows: Section 2 defines the current spectrum management practice and spectrum trading. Section 3 describes the present status of telecom sector of India and Section 4 describes possible regulation need to be formulated for adoption of spectrum trading in India. Section 5 describes about the spectrum requirement for 5G services and Section 6 concludes the paper.

2 Spectrum Trading

The traditional method used for allocation of spectrum is known as 'Command and Control' method [4] as shown in Figure 1. In this method, radio spectrum is divided into different spectrum bands that are allocated to specific radio communications services, such as mobile, fixed, broadcast, fixed satellite and mobile satellite services on an exclusive basis. This command-and-control-based spectrum management framework guarantees that the radio frequency

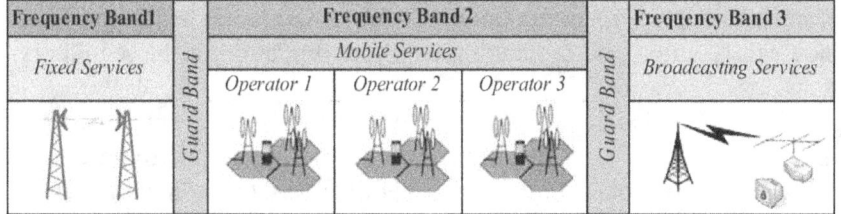

Figure 1 Command and control method of spectrum allocation.

spectrum will be exclusively licensed to an authorized user (i.e. Licensee) and can use the spectrum without any interference.

The command and control method is widely criticized for non-efficient use of spectrum but it has several advantages like providing interference free environment due to rigid technical conditions, provision of guard band between the two allocations and separate spectrum allocation for each service and users. It plays an effective role in the expansion of many services (like GSM) worldwide through co-ordination and harmonization.

The major flaw with the Command and control method is that it does not ensure efficient use of spectrum. Once the spectrum has been assigned to a user, spectrum utilization cannot be questioned during the licensing period provided he fulfill the terms and conditions of the license. Spectrum allocated to one radio communication service cannot be replaced by other services even knowing that spectrum is underutilized. For example, terrestrial TV services, where at least 8 MHz of spectrum (Bandwidth of one channel) remains largely unused but the licensing condition does not permit broadcasters to use this unused spectrum for other services. Another example is mobile telecom services, where a block of spectrum assigned for the entire service area. While spectrum could be heavily used in urban areas, it would be underutilized in rural areas. The operator cannot share this spectrum with other operators under the existing regime.

The current spectrum management approach is under pressure as it does not lead to efficient spectrum use. However, it could be justified in some way that it successfully avoids interference and played an effective role in expansion of few services through co-ordination and harmonization. The emergence of new technologies and services, like cognitive radio, UWB, Broadband, Wi-Max, 4 G etc., need for greater mobility, greater capability of market players and blurring the boundaries between different services and technologies has put a great pressure on radio frequency spectrum manager for efficient and economical use of spectrum. Regulators are adopting different means to improve the spectrum efficiency. Spectrum trading is one tool, through which spectrum efficiency can be improved by making slight modification in the regulatory provisions.

Spectrum trading is a market based mechanism where buyers and sellers determine the assignments of spectrum and its uses [5]. In spectrum trading, seller transfers the right of spectrum usage, in full or part, to buyer while retaining the ownership. Spectrum trading is considered as economically efficient because trade will only take place if the spectrum is worth more to the new user than former it was in the old user [2]. Trading is effective only when it is clubbed with liberalization. Combined have potential to remove the artificial scarcity of the spectrum as trading help to acquire spectrum more readily from the market and liberalization facilitates new technologies/ services to enter in the market. These factors provide consumers with greater choice. The unit of trading can be licensing parameters namely the bandwidth, the geographical area and the duration. This leads to the possibility of partial transfer usage right in terms of bandwidth, area or time. Trading is technically considered as part of a spectrum sharing. However, trading differs in terms of usage right, where the total usage right transferred to seller for a specified period whereas in sharing, seller gets temporary right of spectrum usage, exclusive rights rests with the seller [6].

Trading has first been implemented in New Zealand and Australia. Thereafter, USA, UK, El Salvador, Guatemala and many European countries have introduced spectrum trading in specific bands, which are in demand for commercial use. In Europe, Denmark was the first country, which allowed trading in 1997 followed by Switzerland in 1998. The salient features of spectrum trading in Denmark are given below [7]:

- Spectrum trading is allowed in all GSM bands, 2.1 GHz, 3.5 GHz, 10.5 GHz, 26–29 GHz bands. Spectrum in these bands has been assigned through market mechanisms, auction or public tender
- Default duration for the granted usage is 15 years
- Usage rights can be traded partially for bandwidth and geographical area. No partial trade is allowed for time.
- Reference of usage right (frequency licence number) that will be transferred is required at notification of intent to trade
- Information about the license namely license number, frequency, geographical area etc. Needs to publish prior to the transaction
- Approval of transaction by the Authority is required for the trading part of the licence issued after an auction

- Transaction is required to be made public, which contains information like, identity of parties, reference of original authorization and new authorization, date at which transfer becomes effective, geographic area, frequency band and expiration date.
- Trade may be refused there is unpaid payment for a licence issued following an auction, licence holder cannot guarantee obligations on the provision of services with remaining spectrum.

Spectrum trading has been allowed long back in most countries but the market has not been successful in Australia, New Zealand and European countries because operators in these countries hold good amount of spectrum in each band, and lack of sufficient participation. The activity level is very high in the US where mobile operators have bought spectrum from each other as well as from broadcasters and other niche spectrum holders [8].

3 Indian Telecom Sector

The Indian telecom journey began in 1994, when telecom sector was opened for private sector. The first private sector wire-line and cellular licenses were issued in 1995. From then on, Indian telecom has seen several milestones crossed and many missteps that provided valuable lessons. The telecom industry has witnessed a phenomenal growth in the last decade with 898.02 million phone connections with overall teledensity of 73.32 at the end of March, 2013 (Figure 2), Today, India is the second largest and fastest growing telecom market in the world in terms of the number of wireless connections. The growth is predominately in the wireless sector, which contributes 96.64% (867.80 million) of total phone connections [9].

Telecom services are operating in 800 MHz (CDMA), 900 MHz/1800 MHz (GSM), 2.1 GHz (3G) and 2.3 GHz/2.5 GHz (BWA) in India [3]. India is geographically 5th biggest country with 3.23 million Km^2. The country has been divided into 22 service areas for the purpose of telecom services. These service areas have further been divided into various categories viz. 'Metro', 'A','B' and 'C' based on the population, traffic pattern and revenue generations capacity. Service providers have been given a license for providing telecom services in these service areas. The Indian telecom scenario is bit unusual, if not unique, in terms of the number of operators and assigned spectrum. India is one of the few countries where CDMA and GSM both the technologies are in operation for mobile telecom services. Presently, 15 operators are providing different telecom services in India, out of which

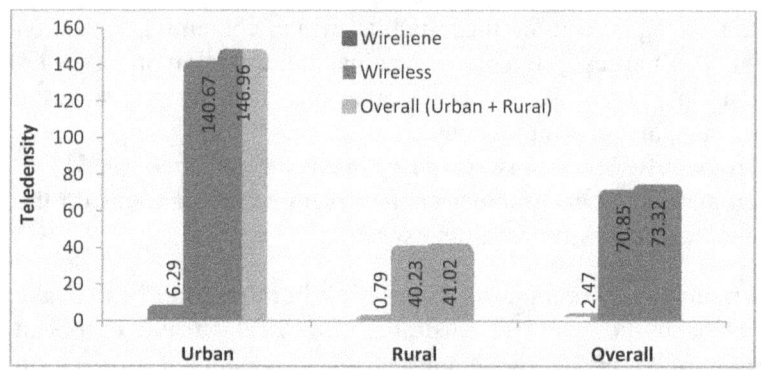

Figure 2 Teledensity in urban and rural india.

8 are providing in all TSAs (6 in GSM and 1 each in CDMA & BWA). The average spectrum hold by each operator is between 10–15 MHz across all bands (i.e. 800, 900, 1800, 2100 MHz paired bands), about one-fourth of the international average [9,10]. A total of 310 MHz across the bands has been allocated for telecom services. There are no pan India presence in CDMA and 3G services except Public Sector Unit (PSU) Bharat Sanchar Nigam Limited (BSNL) /Mahanagar Telephone Nigam Limited (MTNL). BSNL is operating in all TSAs except Delhi and Mumbai, where MTNL is operating. Combinedly MTNL and BSNL present in all TSAs as one PSU. A detail of no. of operators, average spectrum holding etc. is given in Table 1 (BSNL/MTNL considered as an operator).

Initially spectrum assigned for 2G (GSM/CDMA) services to each operator was 4.4 MHz (GSM) /2.5 MHz (CDMA) to start the service. Further, increment made (up to 10 MHz) in piecemeal subject to meeting of strict subscriber base criteria (devised for allocation of additional spectrum) [11]. Therefore, spectrum assigned to operators is so fragmented that in some cases, it is only 0.2 MHz (one carrier of GSM). Almost 85% of the market has been captured by all India operators and remaining 15% market is with other operators. The market share hold by 2G operators (GSM and CDMA) is given in Figure 3.

Similarly, the average spectrum assigned to pan India operators is more than 6.0 MHz in 900/1800 MHz band and other operators have been assigned between 4.4 MHz to 5.0 MHz in a TSA. An entire spectrum of telecom services have not been not assigned through market mechanisms. A detail of allocation of spectrum in different TSAs to various operators is given in Table 2.

Table 1 Details of no. of operators and average spectrum holding.

Services	No. of operators			Frequency band	Total spectrum (in MHz)	Average allocation to an operator (in MHz)
	Partial presence	Pan India	Total			
2G (CDMA)	4	1	5	800 MHz	20+20	2.5+2.5 to 5.0+5.0
2G (GSM)	5	6	11	900/1800 MHz	80+80	4.4+4.4 to 10.0+10.0
3G	4	1	5	2.1 GHz	25+25	5.0+5.0
BWA	1	2	3	2.3/2.4/2.5/2.6 GHz	60	20.0 (TDD)
Total	*14*	*10*	*24*		*310 MHz*	

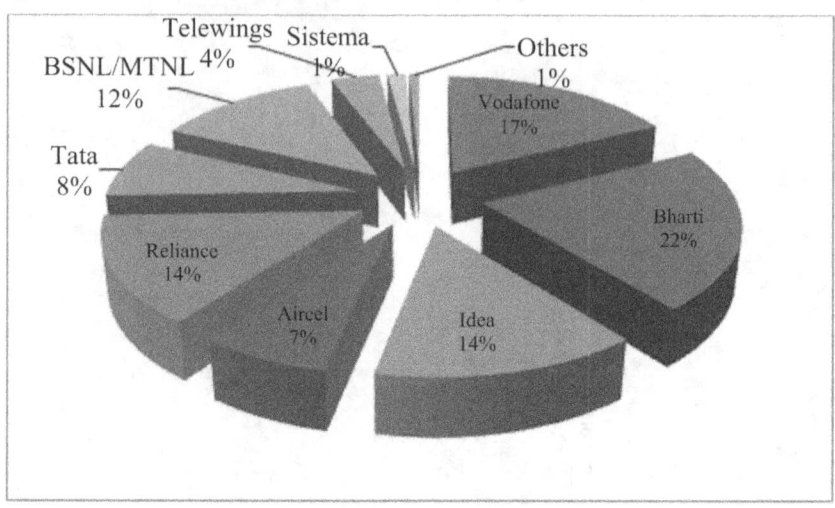

Figure 3   Marker share of 2G operators [9].

The spectrum for 3G and BWA services was auctioned in 2010. In 2012-2013, two phases of the auction was conducted for spectrum in 800/900/1800 MHz band. Almost all assigned 2G spectrum in 800/900/1800 MHz band (except small portion, which sold in 2012 auction in 1800 & 800 MHz band) is through either First Come First Served (FCFS) or Beauty Parade method. Spectrum assigned without auction is technology specific means that they can provide only GSM/CDMA services as applicable in the frequency band. Auctioned spectrum is liberalized and contiguous means that operators are free use any technology [10]. 2G GSM and CDMA services are available almost every part of India. However, no such wide coverage is available in case of 3G services. Roll out of 3G services is limited in urban areas only.

The LTE based 4G service has slowly entered into Indian market in 2.3–2.4 GHz band (BWA-TDD mode). Recently, Bharti has launched 4G service in 4 cities namely Kolkata, Bengaluru, Pune & Chandigarh and likely to be started in Delhi by September 2013 [12]. BSNL has launched WiMAX based 4G services in few cities of India. The remaining players are yet to launch their services. Due to limited coverage and additional capex at this frequency, rollout is limited to metros/urban sectors only. No further rollout is possible in other areas in next 2–3 years, even more time require for covering rural areas. However, if option be given to operators for upgradation of services in 900/1800 MHz band, situation would have been entirely different. Faster rollout is possible due to wide coverage at 900 MHz band. In most of the

Table 2 Spectrum allocation in different TSAs to operators [9].

Service Area	0–5.0 MHz	5.1–6.2 MHz	6.3–8.0 MHz	8.1–10 MHz	Total
Delhi	2	0	1	3	6
Mumbai	4	0	0	4	8
Kolkata	3	1	1	2	7
Maharashtra	4	0	1	3	8
Gujarat	5	2	1	1	9
Andhra Pradesh	4	1	1	2	8
Karnataka	3	1	1	2	7
Tamilnadu	3	0	1	3	7
Kerala	3	1	2	1	7
Punjab	4	1	3	0	8
Haryana	4	2	0	2	8
Utter Pradesh (West)	5	1	1	2	9
Utter Pradesh (East)	5	1	1	2	9
Rajasthan	3	2	1	1	7
Madhya Pradesh	3	1	3	1	8
West Bengal	3	2	1	1	7
Himachal Pradesh	3	3	0	1	7
Bihar	4	1	2	2	9
Orissa	3	1	2	1	7
Assam	1	2	2	1	6
North East	2	2	1	1	6
Jammu & Kashmir	3	1	2	0	6
Total	**74**	**26**	**28**	**36**	**164**

countries, 4G services are being provided in 900/1800 MHz band. The same is not possible in India due to small and fragmented spectrum holdings in 800, 900 and 1800 MHz, license issued for specific technology and uncertainties over the availability of additional spectrum.

The India's telecom sector growth is hampering due to excessive fragmentation of spectrum holdings. The spectrum HHI in India is about 0.15 (in a scale where 1 indicates monopoly occupancy of the spectrum) [8]. Presently, spectrum has been de-linked with the operating license. Any additional spectrum, if required by the operator, will have to buy from the market [13]. Merger & acquisition (between two or more than two companies) is one solution to get additional spectrum but the process is complicated involves coordination with 2–3 different government departments and take considerable time. They can buy additional spectrum through auction but auction can be conducted when free spectrum is available with regulator. Therefore, it is almost impossible for an operator to get additional spectrum in the changed regulatory environment. Spectrum trading could be an option by which an operator can acquire additional spectrum from other operators without going into much complexity.

4 Possible Regulatory Provisions for spectrum trading in India

Spectrum trading has not yet been permitted in India with a view that the market is unbalance, 3–4 operators have captured almost 80% of the market and spectrum assigned to these operators are at nominal price or free of cost (spectrum was bundled with a license) whereas other operators were obtained spectrum through market mechanism [9]. There is a risk of spectrum consolidation which will disturb the market dynamics. Presently, the situation has been changed. The Government de-linked the spectrum from license and announced that spectrum shall be made available at a price determined through market related processes in future. The Government committed in National Telecom Policy 2012 [13] for seamless delivery of converged services in a technology and service neutral environment and to liberalization of spectrum to enable use of spectrum in any band to provide any service in any technology as well as to permit spectrum pooling, sharing and later, trading to enable optimal utilization of spectrum through the appropriate regulatory framework. India could be a good market for spectrum trading due to the following reasons:

- Spectrum hold by any operator is not high in quantum and fragmented
- Number of operators in each service area are considerably high
- Spectrum available through market mechanism only
- Market dynamics is unbalanced; 3–4 operators capture almost 80% of the market
- No pan India operator in 3G services
- Roll out of 3G is not as good as 2G services
- 4G services launched in few cities only
- Tele-density is about at 79

Presently, spectrum holders cannot transfer/lease spectrum. They have two options, either use it or surrender it. No holders can claim spectrum as 'Property Right'. Trading is present form is not possible. Initially permission may be given to holder for transfer/lease of spectrum to materialize trading business. Trading parameters would be bandwidth, location and duration. Besides full transfer, partial transfer of all the three parameters may be allowed as demand in a telecom service area may not be uniform, it is high in the urban sector and low in the rural sector. In case of full transfer, the buyer will have to buy spectrum for the entire service area, which may not be viable from business point. Therefore, location wise (district level) in a TSA may be allowed for trading. A minimum unit for all the three variables should be defined. Unit of location parameter would be district level in a service area, bandwidth may be in multiple of 1 MHz or 0.5 MHz and the unit of duration may be 1 years or its multiple.

4.1 Frequency Band

Market mechanism has so far been adopted for telecom bands only. In the current situation, trading is largely possible in telecom sector only. Auctioned band is the best candidates for spectrum trading. Other bands which fall in the IMT bands category like Public Mobile Radio Trunking System (PMRTS) band (in 800 MHz) may be opened for spectrum trading. Besides access spectrum, operators also need spectrum for backhaul, which assigned in 7 GHz, 14 GHz, 21 GHz and 28 GHz bands. Spectrum in these bands for backhaul services is limited and allotted through FCFS method on administrative pricing basis [3]. Almost 9–10 operators are present in each service area. Backhaul spectrum is very much in demand. It is not possible to fulfill the requirement of operators due to scarcity. These backhaul bands may also be put for trading to mitigate the shortage.

The initial candidate bands for spectrum trading will be: PMRTS band (800 MHz), CDMA band (800 MHz), GSM band (900 MHz), GSM Band (1800 MHz), 3G Band (2.1 GHz), BWA (2.3–2.4 GHz & 2.5–2.6 GHz), BWA band (3.3–3.4 GHz & 3.4–3.6 GHz) and Backhaul bands (7.0 GHz, 14.0 GHz, 21.0 GHz and 28.0 GHz). The spectrum hold by defence, police and other security agencies, public disaster authority and other Government agencies may not be allowed for trading due to security reasons.

4.2 Liberalization of Spectrum

Liberalization is required for efficient usage otherwise objective will be forfeited. Only liberalized spectrum should be allowed for spectrum trading. Auctioned spectrum has already been liberalized while administrative allocated bands are un-liberalized. The Government has made provision that un-liberalized spectrum in 800/900/1800 MHz band can be converted into liberalized by paying one time spectrum charges. However, in other proposed bands, price mechanism may be devised for conversion into liberalized spectrum.

4.3 Roll Out Obligations

As per present spectrum policy, assigned is associated with certain obligations on the part of the license like roll out obligations. There are pre-defined penalties for delay, if roll out obligations are not completed in specified time limit. Regarding licensing obligations, it would be appropriate that trading will be allowed after completion of obligations associated with the license. After transaction of trading, obligations will be lying with buyers for the sold portion of spectrum and obligations corresponding to unsold spectrum will still lying with the seller. A heavy penalty may be imposed on seller in case of non-completion of obligations. This would give a balanced approach towards obligations.

To maintain the market dynamics, a spectrum cap in each band and across all the bands may also be imposed otherwise consolidation may disturb the market harmony. Presently, Spectrum cap has already been imposed while auction of 2G spectrum in 2012 which states the spectrum cap with a band and across all the bands is 50% and 25% of total assigned spectrum respectively. The same may be retained as spectrum hold by an operator is not very high [10]. Besides spectrum cap, a relatively higher annual spectrum usage charges may also be imposed to discourage spectrum hording.

5 Spectrum requirement for 5G

Today's cellular operation is limited between 700 MHz and 2.6 GHz. The global spectrum allocation for all cellular technologies does not exceed 780 MHz and that too in different bands [14]. Each band has their own propagation characteristics, which forced the industry to design radio equipment to accommodate all the bands and technologies. As per Report ITU-R M.2078 [15]: 'Estimated spectrum bandwidth requirements for the future development of IMT-2000 and IMT-Advanced', the predicted spectrum requirements by 2020 would be between 1280 to 1720 MHz. This bandwidth has been envisaged considering demand arises through 4G communication. The next future generation would be 5th generation. In [16], the concept of future 5th generation mobile communication has been given as:

$$4G \ \& \ WISDOM \Rightarrow 5G$$

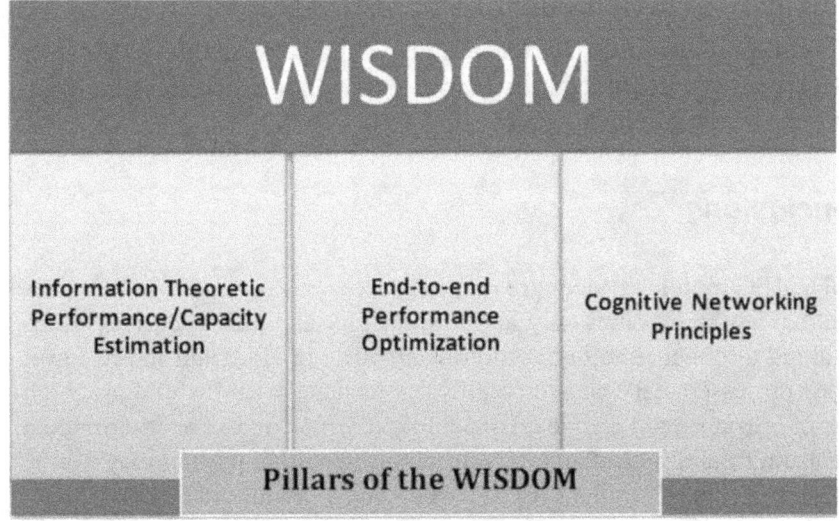

Figure 4 Concept of WISDOM.

The WISDOM, as shown in Figure 4, principle brings unlimited wireless world interconnection, convergence, and cooperation (geographically including cities, countries, continents, and finally, the whole world), together with a large variety of multimedia services at very high data rates, and becomes the main 5G definition point.

Gilder's Law [17] predicts a sixfold increase of the available Bit Rate every 1.5 years it will take approx. six years to achieve the 1Tbps. Therefore, 5G envisage data speed of 1Tbps. The spectrum bands do not have the capability to hold such enormous data. We would have to look at millimetre wave. Recent studies suggest that mm-wave frequencies could be used for wireless communication and it could be augmented the currently saturated 700 MHz to 2.6 GHz bands for wireless communications [14]. Mm-wave spectrum would allow significantly expanding the channel bandwidths and could carry huge data successfully. The future system would lead to a new type of architecture in which connectivity between base stations to user would be provided through millimeter waves. As per NFAP 2011 [3], millimeters wave has been allocated largely for earth exploration satellite and space research. Fixed and Mobile services have also been present as primary services. Spectrum is largely available for research. Therefore, we hope that the vision of 5G could be realized in the near future, which would make a wonderful difference in the peoples' lives. Spectrum trading at millimeter waves may not be applicable at this stage but in future trading may require when congestion would increase. However, a coordination/co-existence study would be required with existing users because services presently working in this range need protection from high power mobile communication.

6 Conclusions

The 2G and 3G mobile services are running successfully in India. 4G services is limited to few metro cities only, and further expansion of 4G services seems to be taking a longer time due to non-availability of spectrum in the lower bands. As discussed in the chapter, highly fragmented spectrum holdings, high number of operators and high level of activity offers a conducive environment for spectrum trading in Indian telecom sector. Spectrum trading may not be much successful in other countries. If implemented in India, it would improve technical as well as economic efficiency of spectrum.

References

[1] Rysavy research. *Mobile broadband Explosion.* www.4gamericas.org/documents/ 4G%20Americas %20Mobile%20Broadband%20Explosion %20August%2020121.pdf
[2] Recommendations on *Spectrum Management and Licensing Framework* (2010). www. trai.gov.in
[3] National Frequency Allocation Plan (NFAP) 2011 of India. www.wpc.dot.gov.in
[4] Martin Cave, Chris Doyle and William Webb. *Essentials of Modern Radio Spectrum Management.*Cambridge University Press (2007).
[5] Carlos E. Caicedo, Martin Weiss. *Characterization and Modelling of Spectrum Trading markets.* http://www.academia.edu/571899/ Characterization_and_Model ing_of_Spectrum_Trading_Markets
[6] Consultation paper *on Overall Spectrum Management and review of license terms and conditions(2009).*www.trai.gov.in
[7] European practices in trading of spectrum usage rights. www.cept.org/.../European% 20practices%20in%20trading%20of%20S
[8] V. Sridhar, Rohit Prasad. Nothing wrong with spectrum sharing. http://www.thehindu businessline.com/opinion/nothing-wrong-with-spectrum-sharing/article4992619.ece
[9] The Indian Telecom Services Performance Indicators March 2013. www.trai.gov.in
[10] Auction of Spectrum. www.dot.gov.in
[11] Spectrum allocation to 2G, 3G & BWA services. www.wpc.dot.gov.in
[12] http://www.airtel.in/forme/wireless-internet/4g_lte/4g_lte_coverage/4g_coverage
[13] National Telecom Policy 2012 of India. www.dot.gov.in
[14] Rappaport T.S *et.al. Millimeter Wave Mobile Communications for 5G Cellular: It will work.* Open Access, IEEE.(2013). DOI: 10.1109/ACCESS.2013.2260813

[15] *ITU-R M.2078 Report.* http://www.itu.int/dms_pub/itu-r/opb/rep/R-REP-M.2078-2006-PDF-E.pdf
[16] Prasad R. Introducing 5G Standardisation. 11thGISFI StandardisationSeries Meeting. www.gisfi.org
[17] Glider's Law. http://en.wikipedia.org/wiki/George_Gilder

Biographies

Purnendu S M Tripathi an Indian Radio Regulatory Services (IRRS) officer of 1998 batch. He is having technical experience of 12 years in the field of Radio Communications/Radio Spectrum Management. He is Engineer in Wireless Planning & Coordination Wing of Department of Telecommunications (DOT), Ministry of Communications & IT, Government of India. In DOT, he is associated with spectrum management activities, including in spectrum planning and engineering and policy regarding regulatory affairs for new technologies and related research & development activities and ITU-R related matter. He was also associated with implementation of a very prestigious World Bank Project on National Radio Spectrum Management and Monitoring System (NRSMMS). His area of interest includes Radio Regulatory affairs for new technologies and cognitive radio. He has also worked as research fellow in Department of Electronics, University of Tor Vergata, Rome, Italy. Presently, he is doing Ph.D. from Aalborg University, Denmark under Prof.Ramjee Prasad.

Ramjee Prasad Ramjee Prasad is currently the Director of the Center for TeleInfrastruktur (CTIF) at Aalborg University, Denmark and Professor, Wireless Information Multimedia Communication Chair.

Ramjee Prasad is the Founding Chairman of the Global ICT Standardization Forum for India (GISFI: www.gisfi.org) established in 2009. GISFI has the purpose of increasing of the collaboration between European, Indian, Japanese, North-American and other worldwide standardization activities in the area of Information and Communication Technology (ICT) and related application areas. He was the

Founding Chairman of the HERMES Partnership – a network of leading independent European research centres established in 1997, of which he is now the Honorary Chair.

He is the founding editor-in-chief of the Springer International Journal on Wireless Personal Communications. He is a member of the editorial board of other renowned international journals including those of River Publishers. Ramjee Prasad is a member of the Steering, Advisory, and Technical Program committees of many renowned annual international conferences including Wireless Personal Multimedia Communications Symposium (WPMC) and Wireless VITAE. He is a Fellow of the Institute of Electrical and Electronic Engineers (IEEE), USA, the Institution of Electronics and Telecommunications Engineers (IETE), India, the Institution of Engineering and Technology (IET), UK, and a member of the Netherlands Electronics and Radio Society (NERG), and the Danish Engineering Society (IDA). He is a Knight ("Ridder") of the Order of Dannebrog (2010), a distinguished award by the Queen of Denmark.

ICT Security and Security Testing Aspects

Mayur Dave[1] and Anand R. Prasad[2]

[1] *mayur215in@yahoo.co.in*
[2] *NEC Corporation, Japan; email: anand@bq.jp.nec.com*

Received July 2013; Accepted August 2013

Abstract

This paper provides an overview of the activities of Security & Privacy Working Group (S&P WG) of the Global ICT Standardisation Forum of India (GISFI). Today telecommunications has become "the utility" of human society and thus is the life-line for not only citizens of India but also for the economy. Therefore there has been serious consideration for security and privacy in and of Indian telecommunication systems and networks. Therefore it was found essential to create a security and privacy working group (S&P WG) in GISFI that would develop solutions for ICT security issues in India as well as provide security solutions for activities in other WG. This paper presents details of the scope, current work accomplishments and future strategy and plans of the S&P WG [1].

Keywords: Telecom Security, Network Element testing, Network testing, methods.

1 Introduction

Since there have been incidents of security breaches in network systems around the world in recent times, the topic of security and consequently the effort of securing network systems has gained momentum.

Journal of ICT Standardization, Vol. 1, 175–186.
doi:10.13052/jicts2245-800X.12a4

Considering this eventuality, as a safety precaution, the Department of Telecom (DoT), Government of India (GoI), has mandated network equipment manufacturers to comply with their latest regulation of having network equipment products tested and certified for security, in Indian labs accredited by the Government [2]. At the same time, the Government has mandated Telecom Licensees (Service Providers) to have their networks tested and certified for security, through periodic network security audits [2]. GISFI S&P WG has undertaken the activity to assist the Indian Telecom industry stakeholders (the Government, the Telecom Service Providers and Network Equipment Manufacturers) by means of preparing and proposing the Telecom Security Framework Proposal, and also Security Testing Methods for Information and Communication Technology (ICT) products and Systems. GISFI S&P WG with its security expertise is analyzing DoT requirements [2] to help in defining and resolving the gaps in the guidelines to enable timely and appropriate implementation of security guidelines.

The paper starts with a section on background and scope to the S&P WG in Section 2 followed by achievements of the group in Section 3. In Section 4 we briefly discuss future plans of the WG and in Section 5 we conclude the paper.

2 Background and Scope

2.1 Background Information

The Security & Privacy Working Group (S&P WG) started as a topic within the Special Interest Group (SIG) and matured with time as a WG. In its short life-span the S&P WG of GISFI has clearly described its role and is working with standardization bodies worldwide. Currently the prime topic that the S&P WG is working on is regarding the Indian network testing requirements. So far, the organizations that have contributed to this WG are NEC, Niksun, TCS, Tata Teleservices, Ericsson, Huawei and I2TB. The working group also had workshops and meetings with participants from DoT, IISc, Tata Teleservices, Airtel, Aircel, MTS and Uninor.

2.2 Scope and Objectives

In this section we present the scope of S&P WG as given in [1].

Security and privacy is of utmost importance in today's industry particularly for Information and Communication Technology (ICT). Work on security must be started from the beginning with complete system consideration otherwise the result is a system full of security holes requiring patch-work. Thus work on security and privacy is required for all technical activities of GISFI. Working on security as a side topic will not suffice therefore a separate technical committee that focuses on security and privacy aspects and has expertise in the field is necessary.

The security and privacy working group performs the following tasks:

1. Study security and privacy including legal intercept requirements regarding ICT for the India.
2. Develop recommendations regarding security and privacy including legal intercept in/for India. This can also include recommendation on algorithms to be used.
3. Perform threat analysis on systems under consideration and technologies being developed by GISFI.
4. Develop security and privacy solutions in collaboration with other committees.
5. Develop legal intercept solutions.
6. Bring Indian requirements on security and privacy including legal intercept to international standardization bodies.

3 Achievements

In this section we give an overview of each of the S&P WG deliverables. These deliverables are the outcome of exhaustive study that we at GISFI S&P WG have performed towards the DoT requirements for security testing mandate to Indian Telecom Network equipment vendors and Network Operators.

The Technical Reports and Specifications have been based on the study of existing Global Wireless Standards and Industry best practices. Recent industry trends and advancements have also been tracked and relevant requirements designed and incorporated while preparing the reports and TSes so as to ensure that the work being done achieves compatibility. These documents have been presented in the GISFI Standardization series meetings and stand approved after review from GISFI members.

A workshop on Indian security requirements between GISFI and DoT was also organized.

GISFI S&P WG is in continuous communication with the Department of Telecom (DoT) and Telecommunications Engineering Center (TEC) of the Government of India. All deliverables of GISFI are submitted to both security section of DoT and TEC.

Table 1 GISFI Security and Privacy WG list of achievements.

Technical Specifications	Significance	Status
Network (and Equipment) Security Requirements and Element Selection Guideline for Security Testing	A comprehensive list of Standards-based Telecom network (element) security requirements, along-with a guideline for assisting on priority-based, selection of network elements to be considered for security testing.	Completed
Telecom Security Framework Proposal for India	Shall present definitions and descriptions for various security testing-related terms, along-with a methodology specified for the each of those terms.	In progress
Technical Reports	**Significance**	**Status**
Telecom Security Framework Proposal for India [3]	Presents definitions and descriptions for various security testing-related terms and information on available methodologies (Standards and Industry Best practices).	Completed
Security in mobile communication systems; Comparison and proposals for India [4]	Presents information on security implementation and/or architecture in the various Cellular technology generations, along-with (potential) security threats in each.	Completed
Security Testing Methods for ICT products [5]	Presents a proposal for security testing methods for security testing for network elements and network (as an entity).	Completed
Element Selection Guideline for Network Equipment Security Testing [6]	A guideline proposal for assisting on priority-based, selection of network elements to be considered for security testing.	Completed
Requirements Study on Circular titled "10-15/2011-AS.III/(21)" [8]	Presents an understanding of the DoT's requirements towards security testing mandate, as mentioned in the said DoT Circular.	In progress
Telecom Security Policy Study and Proposals	Presents an understanding of the DoT's Telecom Security Policy and GISFI proposals towards refining or enhancing the same.	In progress

Considering it as a sign of being appreciated, the work done by the S&P WG has been well accepted by the DoT and parts of it are being used to develop the regulations on security testing in the Indian context.

Our achievements, in terms of documents, are summarized in Table 1.

3.1 Telecom Security Framework Proposal for India [3]

We have presented a technical report outlining the definition and description of ICT security related terminologies as "Network Forensics", "Network Hardening", "Penetration Testing" and "Risk Assessment". The report also presents few standardized methodologies and best practices designed by various organizations for each of the above-listed aspects related to security testing.

Conclusively, we present the topic of "Network Forensics" as the capture, storage and analysis of network events, "Network hardening" as involving the steps taken to secure a network and the devices on it, "Network Penetration Testing" as a test of a network's vulnerabilities by having an authorized individual actually attempt to break into the network in disguise of a hacker and "Risk Assessment" as a process used to identify and evaluate risk and its potential effects, one of the phases of the generic risk management effort.

The report performs such a study on these ICT security related testing procedures, processes and related terminologies which have also been mentioned in the Department of Telecommunications (DoT) (vide Circular "10-15/2011-AS.III/ (21)", dated 31/05/2011) [2]. As a future activity, the report proposes to design a framework, including the same, that will find applicability for testing Mobile Communication Network Equipment's and Systems. Also, we propose on working on solutions leading to actions to fix and prevent recurrence of security problems, as required by the DoT.

A Technical Specification document based on the completed, available Technical Report is in the making.

3.2 Security in mobile communication systems: Comparison and proposals for India [4]

We have presented a technical report on the subject of "Security in Mobile Communication Systems" that is of paramount importance in the times of evolving technologies which is challenged by threats and attacks on them. It provides an overview of the security implementations or architectures provided in the various generations of wireless/mobile communication technologies. The report also provides a brief description of the various security issues in the various phases of technology evolutions in the various generations of wireless/mobile communication technologies. These security issues are in the form of False Base Transceiver Station (BTS) attacks, Interception, Denial of Service (DoS), Distributed Denial of Service (DDoS), Cryptanalysis of algorithms, etc.

As a future activity, we propose a gap analysis between the security mechanisms or architectures in each of the wireless technology generation, for the 3GPP and 3GPP2 technology Standards family, and the weaknesses or security flaws in the same. This would enable network equipment vendors and operators to address existing known issues through firmware or software patch fixes and upgrades.

Further, in this report we have enlisted the Mandatory and Optional requirements (including those for Lawful Interception), regarding the implementation and/or use of security features, as specified in the 3GPP 33-series Technical Specifications (TSs).

This list provides an overview to the concerned Indian Government departments on the 3GPP Standards-based security features, both mandatory and optional, that are to be considered by vendors and operators as part of the Indian Network Security requirements.

Further, this report assists network equipment vendors, network equipment (security) test labs and related stake-holders to select network elements, based on assigned (perceived) priority levels labelled by GISFI, to conduct security testing for network elements according to the DoT's Circular [2] to the TSP's.

The list of network elements, along-with network interfaces and reference points, are derived from the 3rd Generation Partnership Project (3GPP) and 3rd Generation Partnership Project–2 (3GPP2) Wireless Communications Standards.

A Technical Specification document based on the available Technical Report, with the Mandatory and Optional requirements (including those for Lawful Interception) has been presented at the GISFI Standardization Series Meeting #14 and is currently under review by GISFI members.

3.3 Security Testing Methods for ICT products [5]

We have presented a report capturing information about already available security test methods being employed by product/system certification bodies. Examples of such product/system certification schemes are Common Criteria (CC) (testing), and those based on the National Institute of Standards and Technology (NIST) Special Publication Guideline (SP800-115), Open Web Application Security Project (OWASP) Testing Guide (Ver. 3), etc.

The report presents a proposal on 'Network Element Testing' steps in regards to performing security tests on network elements and to certifying them as 'approved/tested/etc. for security' before they are integrated into the mobile network. The proposal outlines a phase-wise testing of network elements, first against Wireless Standards by the 3GPP or 3GPP2 and then against developed Security Targets (STs) and Protection Profiles (PPs) for CC testing. The network elements to be tested for security compliance are listed and assigned a perceived priority as documented in [6].

The report also details a proposal on 'Network Testing' in regards to performing security tests on entire mobile networks deployed by Cellular Operators and to certifying them as 'approved/tested/etc. for security'. It proposes testing of three distinctly identifiable sub-systems (Access Network sub-system, Core Network sub-system and Internet Sub-system), first against Wireless Standards by the 3GPP or 3GPP2 and then against the NIST SP800-115 Guidelines. After successful rounds of testing on the sub-systems, a complete network test, first against Wireless Standards by the 3GPP or 3GPP2 and then against the NIST SP800-115 Guidelines is proposed.

In this report, we have listed the technical and policy gaps between the DoT requirements [2] and the actual (detailed) requirements needed to meet those set by the DoT. This shall help GISFI in working towards providing recommendations with an aim to fill those gaps and assist network operators within the country to practically realize the DoT requirements.

A Technical Specification document based on the completed, available Technical Report is in the making.

3.4 Study on the Common Criteria (CC) [7] in the Indian context

We have presented an input document [7] that introduces the Common Criteria that is a popularly adapted standard for evaluating Information and Communications Technology (ICT) products. This report presents the various technical and procedural aspects related to CC testing and certification. It describes how the various parts (Standards) that collectively make up the Common Criteria have been designed and organized and provides an overview of the contents of each of those Standards. Precisely, the CC has been adopted and standardized by the ISO/IEC into three parts under the ISO/IEC 15408 family (as Part 1, Part 2 and Part 3). Along-with the ISO/IEC 18045 Standard (known as the Common Evaluation Methodology), the ISO/IEC specifies the actions to carry out in the process of evaluation of an ICT product as required by ISO/IEC 15408.

The Common Criteria (CC) is a standardized framework for evaluating ICT products against two types of requirements:

- Security Functional Requirements
- Security Assurance requirements

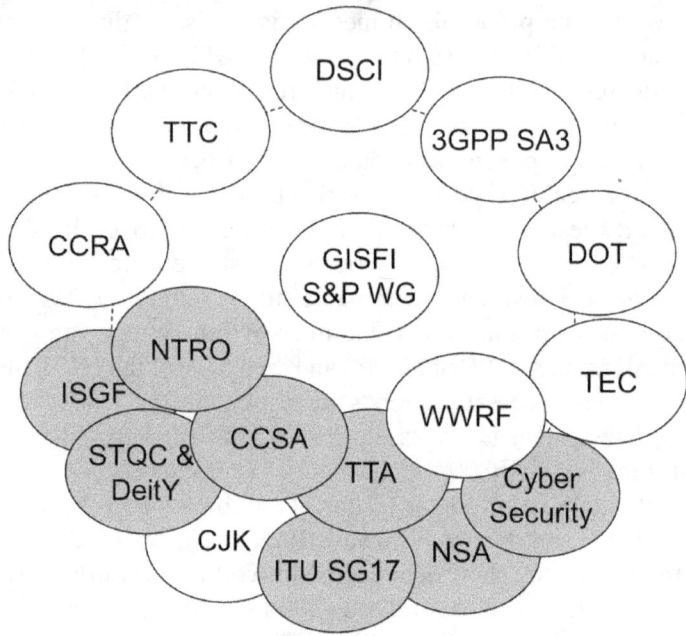

Figure 1 GISFI S&P WG liaison (gray: planned to send liaison).

This report also discusses about the Common Criteria Recognition Agreement (CCRA) and various aspects of certification tasks that the CCRA carries out in co-ordination with various nations of the world.

As mentioned in Section 3.3 and as depicted in Sections 4 and 5, GISFI is actively seeking to engage in and develop towards CC testing and certification for ICT products in the Indian context.

4 WG Plan

Goal of the S&P WG is to have its specifications accepted by the Indian government. For this purpose, strengthening the relation with DoT and, if needed, with other government agencies will be necessary. The TRs of S&P WG is now completed or close to completion therefore activity on technical specifications (TSes) is on-going.

Planned topics, till date, towards developing TSes for assisting TSP's to fulfil the DoT mandate of Security Testing are as follows:

- TS on Network Element Security Testing
- TS on Network Security Testing

- TS on Tools for (Network Element/ Network) Security Testing
- TS on Security Requirements

The WG also plans to develop relationship with several international organizations working on security, some of this is shown in Figure 1.

5 Conclusions

The GISFI security & privacy working group has already produced 1 technical Specification and 5 technical reports towards Indian security requirements. These reports are communicated to the Indian government department of telecommunications. The WG works with Indian mobile operators, vendors and the government thus giving results that balance all needs while fulfilling the security requirements. In the short span of its activity, the S&P WG has also developed relationship with international bodies, for example, 3GPP, CJK, TTC etc. Several new topics are also being developed by GISFI members thus many more results should be accepted from the WG in nea future.

References

[1] GISFI_SIG_201109129, "Security WG Proposal", September 2011.
[2] DoT Circular "10-15/2011-AS.III/(21)", 31 May 2011. URL: www.dot.gov.in/AS-III/2011/as-iii.pdf.
[3] GISFI TR SP.101, "Telecom Security Framework Proposal for India"; December 2012.
[4] GISFI TR SP.103, "Security in mobile communication systems: Comparison and proposals for India"; January 2013.
[5] GISFI TR SP.105, "Security Testing Methods for ICT products"; January 2013.
[6] GISFI TR SP.106, "Element Selection Guideline for Network Equipment Security Testing"; January 2013.
[7] GISFI_SP_201206260, "Report on Common Criteria (CC) in the Indian context"; June 2012.
[8] GISFI TR SP.100, "Requirements Study on Circular titled "10-15/2011-AS.III/(21)", dated 31/05/2011"; January 2013.

Biographies

Mayur R. Dave, MBA- Marketing (Amity University), preceded by a Bachelor of Engineering (B.E.) degree in the field of Electronics and Telecommunications (Mumbai University), is a Senior Manager with an Indian Telecom Operator and has a rich experience working with globally, leading telecommunication product companies in their Research and Development (R&D) divisions for their respective India centers. His expertise is built on testing and validation of telecommunication (end-user) devices to ensure quality and thereby to assist in building and maintaining brand equity for companies as Nokia Mobiles and LG Mobiles that he has worked with. For over a decade, Mayur has played a vital role in formulating quality R&D test processes for organizations and has also contributed to integrating technical superiority into the same through his understanding of global wireless Standards. In his current stint with an Indian Telecom Operator, he is responsible for leading the validation of, and enabling the launch of, quality telecommunication end-user devices for the Operator. As an independent Telecom professional, he also contributes towards GISFI Work Group (WG) research activities in the form of numerous input documents and Technical Reports/Specifications (TR's/TS's), for the 'Security and Privacy' and 'Future Radio Networks' WGs, under the proficient guidance of Dr. Anand R. Prasad.

Vision: Telecommunications and its services will become a part of life as is breathing to humankind.
http://www.prasad.bz

Anand R. Prasad, Dr. & Ir. (MScEngg), Delft University of Technology, The Netherlands, Certified Information Systems Security Professional (CISSP), Fellow IETE and Senior Member IEEE, is a NEC Certified Professional (NCP) and works as a Senior Expert at NEC Corporation, Japan, where he leads the mobile communications related security activity. Anand is a vice-chairman of 3GPP SA3 (mobile communications security standardization group). He is a Member of the Governing Body of Global ICT Standardisation Forum for India (GISFI) where he is founder chairman of the Security & Privacy working group and was chairman of the Green ICT working group. Before joining NEC, Anand led the network security team in DoCoMo Euro-Labs, Munich, Germany, as a manager. He started his career at Uniden Corporation, Tokyo, Japan, as a researcher developing embedded solutions, such as medium access control (MAC) and automatic repeat request (ARQ) schemes for wireless local area network (WLAN) product, and later he was a project leader of the software modem team. Subsequently, he was a systems architect (as distinguished member of technical staff) for IEEE 802.11 based WLANs (WaveLAN and ORiNOCO) in Lucent Technologies, Nieuwegein, The Netherlands, during which period he was also a voting member of IEEE 802.11. After Lucent, Anand joined Genista Corporation, Tokyo, Japan, as a technical director with focus on perceptual QoS. Anand has provided business and technical consultancy to start-ups, started an offshore development center based on his concept of cost effective outsourcing models and is involved in business development.

Anand has applied for more than 50 patents, has published 6 books and authored more than 50 peer reviewed papers in international journals and conferences. His latest book is on "Security in Next Generation Mobile Networks: SAE/LTE and WiMAX", published by River Publishers, August 2011. He is a series editor for standardization book series and editor-in-chief of the Journal of ICT Standardisation published by River Publishers, an Associate Editor of IEEK (Institute of Electronics Engineers of Korea) Transactions on Smart Processing & Computing (SPC), advisor to Journal of Cyber Security and Mobility, and chair / committee member of several international activities. He is a recipient of the 2012 (ISC)² Asia Pacific Information Security Leadership Achievements (ISLA) Award as a Senior Information Security Professional.

Convergence through Next Generation Cloud and Service Oriented Networks in the Indian Scenario

Parag Pruthi[1], Ashutosh Dutta[2], Niranth Amogh[3]
and Ritesh Kumar Kalle[4]

[1] *NIKSUN USA, email: parag@niksun.com*
[2]*AT&T USA, email: ad5939@att.com*
[3] *Huawei R&D India, email: namogh@huawei.com*
[4] *NEC India, email: ritesh.kumar@necindia.com*

Received July 2013; Accepted August 2013

Abstract

Cloud and Service Oriented Networks are integral part of the convergence strategy laid out by Government of India through the National Telecom Policy 2012. This paper describes the key requirements for converged cloud and services infrastructure framework and architecture that have impact on standardization. The requirements not only include perspectives of Indian challenges for the infrastructure but also include the challenges imposed by innovative applications and services that are to be delivered through the Next Generation of Cloud and Service Oriented Networks.

Keywords: Cloud, Service Oriented Networks, Applications, Service Infrastructure, Convergence, Service Delivery, Standardization, Context Awareness, Self Organization, Emergency Services.

1 Introduction

Service oriented network principles provide the necessary foundation for delivering the vision of convergence of service delivery for multiple market segments. Cloud computing is adopted to build cost and energy efficient

Journal of ICT Standardization, Vol. 1, 187–204.
doi:10.13052/jicts2245-800X.12a5

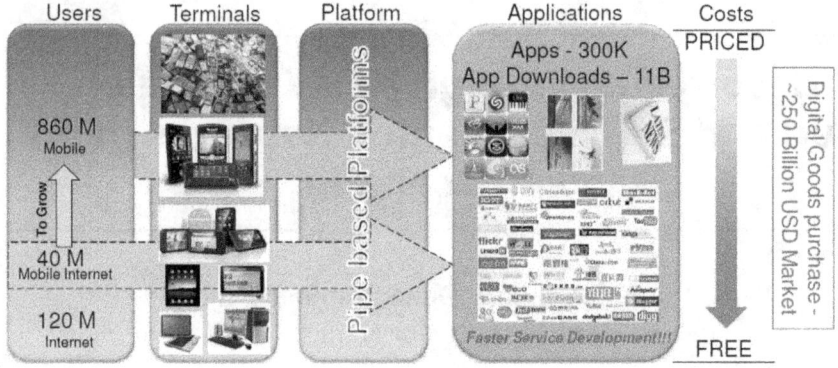

Figure 1 Indian market scenario.

service delivery mechanisms. Most recently the Indian government has laid out its objectives and missions in the area of convergence and cloud services through the NTP (National Telecom Policy) 2012 and the Draft National Policy on Information Technology 2011. Cloud & Service Oriented Networks are intended to support multimodal communication environments where information can be communicated through a variety of terminal devices, network access technologies, and underlying infrastructures. The information may be presented in real-time (e.g., interactive voice) or time-shifted (e.g., voice mail), in its original format (e.g., analog speech) or transformed (e.g., file attachment). The information can be delivered by the network to a location, a device, or a person, reflecting personal preferences and mobility options.

Figure 1 shows that the number of mobile subscribers in India is increasing at a high rate and is only next to China. Even the mobile Internet growth is increasing steadily owing to the proliferation of end user devices which deliver various levels of user experience. The application space has been struck by various innovations quite rapidly and is now being delivered by Over The Top (OTT) providers using the telecom network. A study from TellLabs suggest that operators or telecom service providers worldwide especially in developing countries need to transform their networks and business models to be aligned to Cloud and Service Oriented Networks in the next 3 to 5 years.

2 Requirements and Gaps for India

Service providers are highly interested in leveraging existing networks and infrastructure to increase the value of those networks by enhancing their

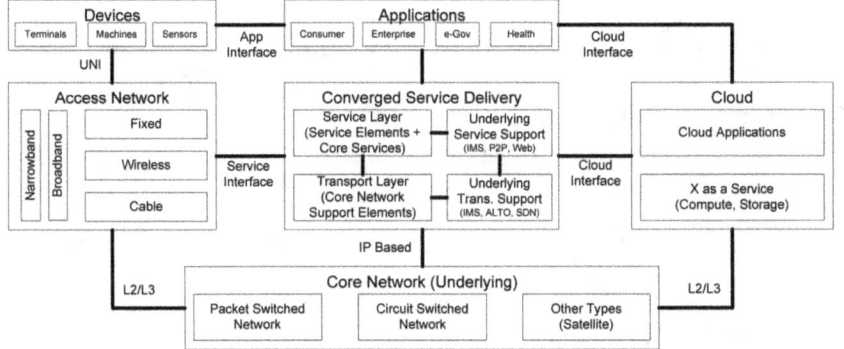

Figure 2 Convergence framework for cloud and service oriented networks.

ability to offer customers "seamless" delivery of applications independent of any access or transport technology. This framework provides a common architecture and set of service interfaces to address this basic goal. Adhering to this architecture and to the services and service models set forth, provides a common framework for delivering services, irrespective of the network context. Regulatory requirements may affect any telecommunications services provided. These requirements can be classified broadly in a framework as shown in the Figure 2.

A service oriented network enables the new providers to offer value added services while using the existing underlying services. It shows a multi-tiered architecture. At the lowest layer there are multiple types of networks providing a layer of network convergence. On the top of network layer there is common service platform layer. This layer consists of services as offered by IMS (IP Multimedia Services) and OMA (Open Mobile Alliances). Current operators and new service providers offer new variety of services using the third party APIs and protocols. CSeON WG will define these APIs and protocols that allow the existing service providers and the new ones to offer a variety of services.

Figure 3 shows how CSeON domain takes the best of three domains such as IMS (IP Multimedia Subsystem), Cloud/SOA (Service Oriented Architecture) and WEB 2.0. While IMS is suitable for Telco, WEB 2.0 is most suited for Web domain. On the other hand, Cloud/SOA is most suited for IT domain.

2.1 Convergence

The CSeON target architecture must not only separate services from transport, but must enable efficient interworking between applications to support

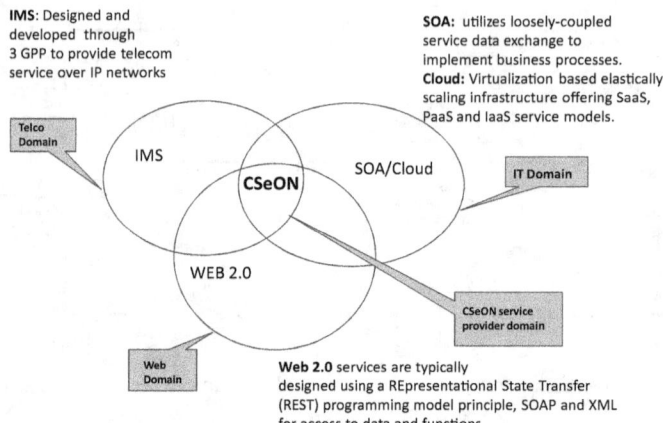

Figure 3 Cloud and service oriented networks domain.

innovative converged services. Many of today's services are vertically-integrated which inhibits integration with other applications. This interworking can add value to the existing as well as new services by integrating them into the larger convergence of media and access modes, allowing service providers to roll out the kind of customized and convenient advanced services that segments of their markets are already asking for today. By putting the disparate parts of the communications puzzle together (such as wireline and wireless services, switched and IP networks, voice and other media, and access modes of all types), service providers have the flexibility to create the right combinations of services for their markets and deploy them to the benefit of end customers.

To make the transition to a fully converged CSeON network, service providers need a standards-based, converged service-enabled network architecture that is ready and able to deliver value-added services. To fulfill these needs, the resulting architecture should support:

- Open, standards-based interfaces allowing "plug-and-play" integration of any number of applications.
- Full convergence of services over a number of access modes — blending instant messaging with unified communications and VoIP, for example.

2.2 Rural India

As per the Census of India,approximately 70% of country's population lives in Rural India. There are almost 600,000+ villages in the country with 1000 people per village and per capita income of Rs. 20 to 25 per day. The rural tele-density in India is 40.07% as compared to urban tele-density of 148.46%, as on January 2013. The country's tele-density is 73.07% compared to world's tele-density of 96%.

The people in rural India, needs connectivity in order to be fully connected to the progressing world in order to make it a truly symbiotic society. For telecom operators, it is thousands of customers subscribing every day while for equipment vendors it brings a unique challenge to develop new solutions that provide the access to these communities.

Following are the requirements specific to rural India,

- Low population density
- Low income levels
- Lower literacy levels
- Sparsely populated and geographically dispersed areas

So, any business model prepared for rural area should be:

- Affordable (cheaper)
- Reliable
- Rich in Contents (like Value Added Services specific to the needs of people working in agricultural domain)
- Available in local language
- Self sustainable
- Replicable
- Scalable
- Easier and faster to deploy
- Deploy alternate source of Energy like Solar power
- Low maintenance and Support Cost

Low maintenance network equipment suitable for rural environment considering the following factors,

- Lack of continuous power supply
- Lack of high available infrastructure
- Harsh environmental conditions
- Low literacy level

2.3 Deliver Secure Applications and Services

Security is one of the most essential enablers and differentiators in CSeON. Identifying services and the users, authorizing the service to access specific

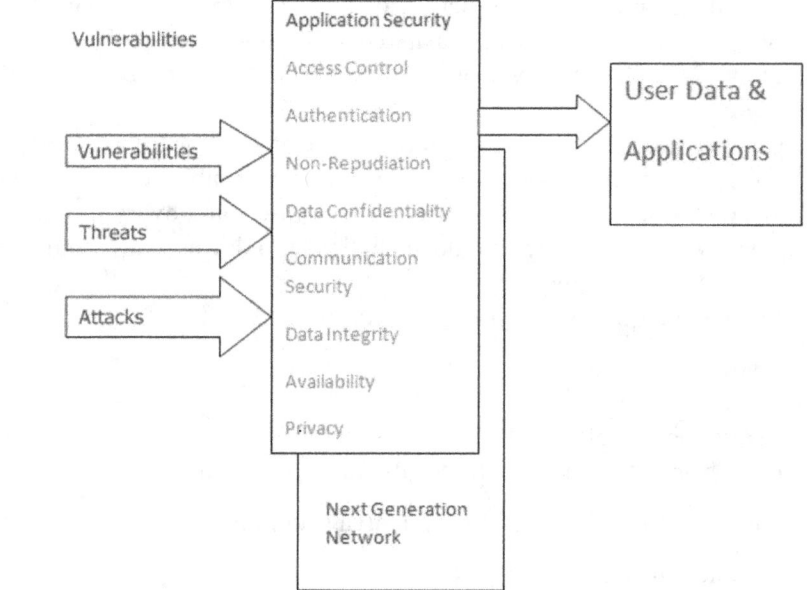

Figure 4   Eight security enablers.

content based on one's access rights and authenticating the service to use appropriate resources are required. The essential enablers are shown in Figure 4.

- **Access Control**: Enabling access authority to control the access of the resources, protecting of copy rights. Implement the policy to describe how to secure access a data resource.
- **Authentication:** Confirming the identity of the person, tracing its origins and ensure that the application is the same as it intends to. Focus is on authentication for both device and end users and how far authentication goes: User to phone (for highly secure environments); User Agent to IMS; IMS to User Agent; User to IP-CAN; IP-CAN to User; User Agent to User Agent; Cross domain authentication; Authentication of Signaling messages; Authentication of media packets; Authentication of messages that traverse PSTN<-> Packet Networks through Signaling/Media gateways.
- **Non-repudiation:** Techniques like Digital signature should be used not only to ensure that a message or document has been electronically signed by the person. Ensure that a person cannot later deny that they furnished the signature.

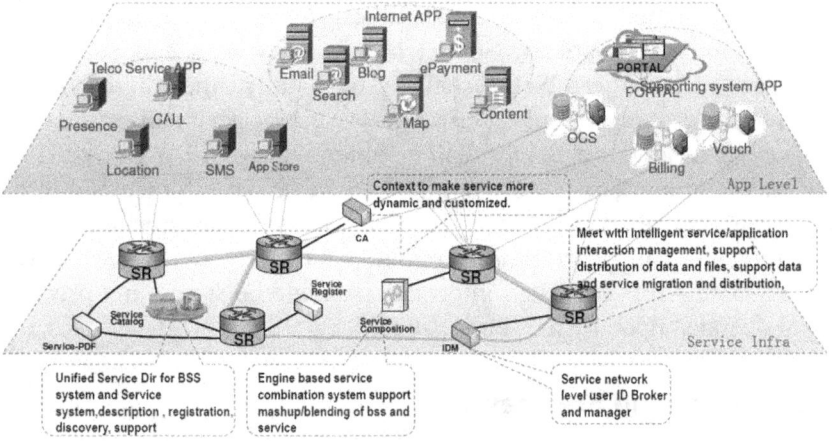

Figure 5 Service cloud.

- **Data confidentiality:** Provide infrastructure for encryption and password authentication.
- **Communication security:** Platform to deny unauthorized access of data and resources. Security refers to the need to ensure that communications between endpoints cannot be unlawfully intercepted or redirected.
- **Data Integrity:** Assurance that data is consistent certified and can be reconciled.
- **Availability:** Resources and data need to be available when required.
- **Privacy:** CSeON can be combined with identity authentication (e.g. RFID can be used).

2.4 Cloud Services

CSeON shall be able to leverage all the service models of Cloud like IaaS, SaaS, PaaS(as described by NIST) to allow an operator to provide and operate a cost efficient CSeON. CSeON should be able to provide a PaaS layer for CSeON services.Shall enable a service eco-system by combining services from different networks and enabling stakeholders to develop innovative applications and deliver seamlessly over any network. For example, a service infrastructure consisting of service processing nodes which enables seamless integration and interaction of services from telecom and Internet as shown in the Figure 5.

- **Support SaaS model:**CSeON shall be able to deliver services like mashup applications as Software as a Service based on pay per use model.
- **Compute as a Service:**CSeON shall be able to utilize the Public/Private cloud for processing complex service tasks such as personalization
- **Storage as a Service:**CSeON shall be able to utilize the public/private storage cloud on demand to meet the high scale data storage and also be able to support the real-time and non-real time data/content access requirements. This public storage requirement is only for non-sensitive data/content. All user-related information should be stored in a private cloud operated upon by the service provider.

2.5 Applications and Services

Indian telecommunication deployment is unique in terms of geography and user density. Currently the wireless subscribers (CDMA, GSM/GPRS) dominate the subscriber base with close to 900 Million subscribers in 23 telecom circles and rest are fixed line subscribers (PSTN, DSL) with a base of around 30 Million. While the wireless space is dominated by Private Telecom Service Providers (TSPs), the wired telecom is still under the domination of Public Sector Undertaking (PSU) telecom companies. Though the urban tele-density has reached close to 160 percent, the rural tele-density is still lagging behind at 40 percent penetration. Indian telecom market has evolved over time and consists of—multiple operators, multi—vendor equipments, multi—technology (GSM, CDMA, 3G, LTE) equipments, wired—line and long distance networks.

Since all these technologies are governed by separate specifications, it is important to find technology solutions and standardize the interconnectivity for various applications and services which will be offered through a converged services framework. CSeON will enable some of the following services through the converged services framework:

2.5.1 Emergency Telecom Services

Since emergency is closely associated with the national security issues, appropriate regulatory requirements and Government policies are to be considered in this regard. The following key India specific requirements are envisioned:

- **Numbering Plan:** All the major categories of emergency services like Police, Fire, Ambulance, Hospital Related, Emergency Information Services, Emergency Disaster Management Services, Police Related Services should be available through a single number. Thus, for standardizing the single number emergency service, the numbering codes need to be defined.

- **Location Tracking and Caller Identification Issues:** Indian operators have currently not installed any form of mobile location tracking systems. DoT has set the deadline of year 2014 to install such a mobile location tracking system that is much more precise than the current system that provides the Cell ID of the subscriber. Centralized database of numbers for tracking handsets across all operators with confidentiality. Low GPS penetration in Indian subscribers implies that the solution needs to depend on triangulation and other similar mechanisms supported by the network infrastructure. Real-time transfer of caller identification & location information is needed. Spare capacity in public networks needs to be reserved for handling of emergency traffic,location tracking for SMS, data calls, SIM-less emergency calls, etc.
- **Public Service Access Points (PSAP):** The PSAP should be accessible for the single number and should be reachable over all telephony networks within the country with dedicated resources provisioned. Both centralized and distributed architectures should be analyzed with controls provided to independent authority.
- **Local Language Support and Social Inclusion:** Indian language diversity is a big challenge to provide good reachability of emergency services. These services should be available in local languages. As part of the social inclusion, the emergency services should be made available to persons with speech and hearing impairments with non-voice accessibility such as text and gesture based solution.
- **National Security Implications:** The national emergency services infrastructure should be protected from various threats such as physical and software related attacks (virus, malware, DoS, etc), network failures (power failure, infrastructure damage, etc), congestion due to network overload during disasters and hoax calls.

2.5.2 E-Health Services

As an initial effort to connect the medical network to the end users, e-Health services should be effectively deployed by CSeON. e-Health services should improve the accessibility of medical services to rural sector in particular. Gradually the government should be able to introduce e-Health applications like electronic health records, almost real-time automated health monitoring and emergency services, diet control and health tips.

2.5.3 E-Governance Services

In the area of e-Governance the applications aiding communication like Government to Citizen (G2C), Government to Business (G2B), Government to Governments (G2G), Government to Employees (G2E) and Citizen to Government (C2G) should be effectively deployed by CSeON by harnessing existing citizen identity projects like Aadhaar.

3 Standardization Approach

The GISFI Cloud & Service Oriented Network (CSeON) Standardization Working Group is driven by the business needs of the Indian market. The goal is to produce CSeON Standards, consistent with the unique Indian regularity, business and infrastructure (urban as well as rural India) requirements. The GISFI CSeON Working Group will focus on providing a phased business-driven action plan for achieving implementable and interoperable CSeON standards.

Following is the list of other standards forums that could collaborate closely with the GISFI CSeON WG or provide ongoing liaisons. These include:

- Ministry of Communication and Information Technology (MIT)
- Cellular Operators Association of India (COAI)
- ITU-T: SG13, including the Focus Group on Next Generation Networks (FGNGN)
- 3rd Generation Partnership Project (3GPP)
- European Telecommunications Standards Institute (ETSI) TISPAN
- IEEE P1903 (NGSON) project
- 3rd Generation Partnership Project #2 (3GPP2)
- GISFI Technical Committees

3.1 Candidate Areas for Standardization in CSeON

- **QoS:** End to End QoS for composed services/applications in a multi-operator environment.
- **Service Composition:** Composition of services/applications in a multi-operator environment,especially rapid application adaptation for semi-urban and rural applications.
- **Self Organization:** Self organization of service operations applicable in a multi-operator environment.

- **Content Delivery:** Content delivery in a multi-operator environment. Efficient mechanisms to reach semi-urban and rural users.
- **Context Awareness:** Adaptation of application based on contexts of users, networks, device and services.
- **Identity Management:** Handling multi-identity in the form of Global ID across multi-operator environment.
- **Cloud security monitoring:** Monitoring the attack vectors to avoid any disruption of cloud services

4 Cloud and Service Oriented Networks WG Charter

The main objective of GISFI CSeON WG is to develop phase-wise standards and implement through suggestive Generic Requirements (GRs) from Government of India. The CSeON WG has focussed its tasks and activities as follows:

- **Analysis and Requirements:** Gather CSeON requirements and establish liaison with Indian stakeholders. Publish technical reports in this subject matter.
- **Standardization:** Develop standards based on the gaps and provide a platform to recognize these Indian requirements and standards in the international standards community. Publish standards in India and also make contributions to international standard bodies.
- **Implementation & Deployment:** Participate with all Indian stakeholders to implement the standards and provide consultation to the Indian government on the deployment models and regulations for a establishing and operating a sustainable business.

4.1 Technical Reports

As part of the analysis activities the following technical reports will be published by the CSeON WG:

- **CSeON Requirements and Framework:** This technical report provides detailed requirements for Cloud and Service Oriented Networks and in particular large focus is laid on Indian requirements.
- **Business Models for CSeON:** This technical report focuses on eliciting the business scenarios and challenges which focus on establishing and operating a long term sustainable business based on CSeON.

- **Information Management for CSeON:** This technical report focuses on technical challenges involved in managing large scale information operated upon by the CSeON providers. It also proposes method to migrate legacy information system to CSeON model.
- **Carrier Grade Cloud Computing:** This technical report provides new and emerging architectures to provide high performance, cost and energy efficient compute, data, platform and software services.
- **Emergency Telecom Services:** This technical report provides technical challenges and standards involved in providing emergency services in Indian context.

4.2 Collaborations

Collaboration is key to building a sustainable environment for innovative standardization. It is important to work together in an ecosystem where similar problems are being solved and fit together in the overall convergence vision. CSeON WG has taken up the collaboration activities with the following groups in other standardization bodies.

- **IEEE P1903 (NGSON):** This group focuses on developing standards in the emerging area of Next Generation Service Overlay Networks (NGSON). CSeON WG has successfully setup liaison with this SDO.
- **IETF CCNS:** This group focuses on developing RFCs for Cloud Computing and Network Services area. A liaison is initiated with IETF Internet Architecture Board (IAB) in this area.
- **TEC:**Telecommunication Engineering Centre is the Indian government body responsible for investigating standards and researching new products and services. A liaison is initiated with Information Technology (I) division and the Next Generation Networks (NGN) division.
- **ATIS CSF:**Cloud Services Forum focuses on developing framework for Cloud Services. A liaison is intended with this group.
- **ETSI NFV:** ETSI Network Function Virtualization group focuses on virtualizing the network functions in next generation networks including both wireline and wireless networks. Efforts are underway to set up collaboration between NFV and IEEE NGSON.

5 Conclusion

The diverse Indian requirements aimed at convergence of networks, services and devices are only possible through standards based interoperable cloud and service oriented networks infrastructure. A strategic convergence framework is required to enable diverse business needs as shown in Figure 2. This convergence framework will be a collaborative effort which involves combining standards from various industry bodies like GISFI, TEC, IEEE, ATIS, ETSI, etc.The goal of GISFI CSeON working group is to collaborate and produce CSeON Standards, consistent with the unique Indian regularity, business and infrastructure (urban as well as rural India) requirements.

References

[1] Government of India, National Telecom Policy – 2012, May 2012, http://www.dot.gov.in/ntp/NTP-06.06.2012-final.pdf

[2] Government of India, Dept. of Electronics and Information Technology, National Policy on Information Technology (NPIT 2011), Oct 2011, http:// deity.gov.in/sites/upload_files/dit/files/National_Policy_on_Information_Technology_07102011(1).pdf

[3] Parag Pruthi, AshutoshDutta, Niranth Amogh, CSeON Requirements and Framework, Draft, Sep 2012, http://gisfi.org/wg_documents/GISFI_CSeON_201209310.doc

[4] Anand R Prasad, Business models for CSeON, Jun 2011, http:// gisfi.org/wg_documents/GISFI_SeON_20110675.doc

[5] ATIS Cloud Services Forum (CSF), 2009, http://www.atis.org/cloud/index.asp

[6] IETF NVO3 (Network Virtualization Overlays), http://datatracker.ietf.org/wg/nvo3/charter/

[7] IETF SDNRG (Software Defined Networking Research Group), http://trac.tools.ietf.org/group/irtf/trac/wiki/sdnrg

[8] SCIM (System for Cross-domain Identity Management), http://datatracker.ietf.org/wg/scim/charter/

[9] IEEE NGSON Whitepaper, 2008, http://grouper.ieee.org/groups/ngson/P1903_2008_0026-White_Paper.pdf

[10] TellLabs Study, Feb 2011, http://www.tellabs.com/news/2011/index.cfm/nr/142.cfm

[11] TRAI: Telecom Subscription Data, Jan 2013, http://www.trai.gov.in/ Write Read Data/WhatsNew/Documents/PR-TSD-Jan2013.pdf

[12] IAMAI: Internet and Mobile Association of India http://www.iamai.in/

[13] The NIST definition of Cloud Computing, 800-145, Sep 2011, http://www.nist.gov/manuscript-publication-search.cfm?pub_id=909616

[14] NIST Cloud Computing Reference Architecture, 500-292, Sep 2011, http://www.nist.gov/manuscript-publication-search.cfm?pub_id=909505

[15] Aadhar enabled Service Delivery, Feb 2012, http://uidai.gov.in/images/authDoc/whitepaper_aadhaarenabledservice_delivery.pdf

[16] Guidelines on Security and Privacy in Public Cloud Computing, Dec 2011, http://www.nist.gov/manuscript-publication-search.cfm?pub_id=909494

[17] The Indian Telecom Services Performance Indicators July - September 2012, http://www.trai.gov.in/WriteReadData/PIRReport/Documents/Indicator %20 Reports %20-% 20Sep _2012.pdf

[18] National number plan http://www.dot.gov.in/numbering_plan/nnp2003.pdf

[19] COAI Response to DoT instruction to setup Location Based Services in mobile networks http://coai.in/docs/PressReleases/Press%20Release%20-%20Location%20Based%20Services%20-%2016-6-2011.pdf

[20] Draft Amendment to the Telecommunications (Emergency Call Service) Determination 2002 – blocking of SIM-less calls, Consultation Paper, ACMA, Australia, 2007. www.acma.gov.au

[21] Mobile Standards Association of India URL: http://www.msai.in/

[22] COAI input at ETS Workshop Organized by TRAI http://www.trai.gov.in/WriteReadData/Events/Presentation/PPT/201211050554241072138Shri%20Vikram%20Tiwathia%20-%20Associate%20Director%20General,%20COAI.pdf

[23] National Emergency Number Association Standards http://www.nena.org/?page=Standards

[24] Expectations of the deaf and hard of hearing communities http://www.911.gov/pdf/TDI-11072006.pdf

[25] Crime and Criminal Tracking System, MHA, Govt. of India http://ncrb.nic.in/cctns.htm

[26] Instructions on verification of new mobile subscribers (Prepaid and Postpaid) http://www.dot.gov.in/as/2012/DOC181012.pdf

[27] Single number access for non emergency services. http://pib.nic.in/newsite/erelease.aspx?relid=86571

[28] Rural Urban Distribution of Population, Census of India 2011. Govt. of India. http://censusindia.gov.in/2011-prov-results/paper2/data_files/india/Rural_Urban_2011.pdf

[29] ICT Facts and Figures 2013, ITU-T http://www.itu.int/en/ITU-D/Statistics/Documents/facts/ICTFactsFigures2013.pdf

Biographies

Dr. Pruthi brings over twenty-five years of expertise in the network security, wireless and applications analysis industry. Dr. Pruthi is the founder of NIKSUN which he has built from a startup to a highly successful global company leading the way in the cyber security, wireless and network monitoring markets.

Dr. Pruthi is widely recognized as the founding father of packet capture, stream to disk, bit vacuum, and other similar technologies which he brought to market in 1997 as NetVCR. He is also widely accepted as one of the leading innovators in the field of cyber security. In 2001 he introduced the NetDetector as the only device in the world capable of multi gigabits per second line rate recording and simultaneous analysis, reconstruction and replay in order to discover the source of security incidents and identification of potential information leakage. This invention led to the creation and growth of the field of network forensics.

Recognized as one of the foremost experts in advanced cyber security technologies, Dr. Pruthi advises on cyber defense strategies with some of the highest levels of governments and enterprises around the world. He is a frequent speaker on cyber security issues, including keynotes to delegates from across 28 NATO nations, agencies, and strategic commands at the NATO Information Assurance Symposium 2010 and 2011.

Dr. Pruthi holds a Bachelor's degree in Electrical Engineering and a Master's in Computer Science from Stevens Institute of Technology. He also has a Doctorate in Telecommunications from The Royal Institute of Technology, Stockholm, Sweden, and his thesis, "An Application of Chaotic Maps to Packet Traffic Modeling," was the first to apply chaotic and fractal or self-similar systems to accurately model the seemingly erratic nature of network traffic.

Dr. Pruthi has received many honors, is interviewed frequently by news and media and asked to write various articles on a diverse set of topics. Notable were his interview with General Norman Schwarzkopf on Cyber Security which aired on CNBC and various other media outlets; cover story "Securing You Against The Unknown" in Silicon India Magazine's September 2011 issue; feature cover story in CIO Review's inaugural issue (April-May 2012); and a chapter on "Delivering a Long-Term Vision for Software" in the

"Inside The Minds" series on "Growth Strategies For Software Companies." In 2012 SmartCEO magazine awarded Dr. Pruthi CEO of one of the BEST-RUN COMPANIES in the Mid-Atlantic. In the same year, Network Products Guide named Dr. Pruthi Outstanding Leader of the Decade in the Information Technology Industry. In addition to a host of awards, his most cherished award is that from his peers at the IEEE who honored him with the IEEE Region 1 Award of Managerial Excellence for his Leadership, Entrepreneurship, and Cybersecurity Vision.

Dr. Ashutosh Dutta currently works as a Lead Member of Technical Staff at AT & T. Earlier he worked as CTO Wireless at NIKSUN Innovation Center in Princeton, New Jersey where he is leading the research and development efforts in the area of 4G networks and service oriented networks. Most recently he worked as a senior scientist and project manager in Telcordia Technology's Applied Research, Piscataway, NJ for 13 years. Prior to joining Telcordia, Ashutosh was the Director of Central Research Facilities at Columbia University, from 1989 to 1997, and worked as a computer engineer with TATAs (Telco), India, from 1985 to 1987. Ashutosh's research interests include wireless Internet, multimedia signaling, mobility management, 4G networks, IMS (IP Multimedia Subsystems), VoIP and session control protocols. Ashutosh contributed to the research community by publishing his research results in journals, conferences, workshops and tutorials and serving as the general chair and technical program chair. He has published more than 80 conference, journal papers and Internet drafts, three book chapters, and has given tutorials in mobility management at various conferences. He was awarded 2009 IEEE MGA Leadership award and 2010 IEEE-USA professional leadership awards. Ashutosh has 19 issued US patents. Ashutosh Dutta is a senior member of the IEEE and the ACM. He obtained M. Phil. and Ph.D. in Electrical Engineering from Columbia University, New York. MS in Computer Science from NJIT and BS in EE from NIT Rourkela, India.

Mr. Niranth Amogh is the Sr. Lead Researcher at Huawei India R&D Center at Bangalore and currently leads the NGSI (Next Generation Service Infrastructure) Technology Introduction Group and is responsible for Research and Standardization activities within the organization. His research interests include NG-SON, SDN, Context Aware Computing, P2P, Cloud/Grid computing, Carrier Grade Cluster and Middleware and Telecom Network Management and he has filed several patents in his research. He is a member of various global telecom standardization bodies and technical societies, including a senior member of IEEE, a member of IEEE-SA, IEEE ComSoc, and ACM, as well as being principal contributor and project officer of the IEEE P1903 (NGSON) standards. Amogh is also the Vice-Chair for the Global ICT Standardization Forum for India (GISFI) Cloud & Service Oriented Networks (CSeON) Working Group focused on standards development in India.

Dr. Ritesh Kumar is Research Engineer at NEC Mobile network Excellence Center (NMEC), Chennai, India. Prior to this, he was the Senior Scientific Officer and a full time PhD scholar in Wireless Networks Lab at IIIT-Bangalore. He primarily works in related areas of modelling and performance analysis of Mobility, Power Saving and Quality of Service in Wireless networks, particularly LTE, WiMAX and Wi-Fi. His research was supported by Microsoft Corporation under the Microsoft Research India PhD Fellowship Award and the Department of Information Technology, Govt. of India. Ritesh has a graduate degree in Information Technology from IIIT-Bangalore. He has applied several patents and has published extensively in peer reviewed International Journals and Conferences.

Green Information and Communication Technology Standards Development: An India Perspective

Ritesh Kumar Kalle[1] and Arvind Mathur[2]

[1]*NEC India Pvt. Ltd., India; email: ritesh.kumar@necindia.in*
[2]*Cisco Systems, India; email: arvmathu@cisco.com*

Received July 2013; Accepted August 2013

Abstract

Green Information and Communication Technologies (GICT) refers to the wide-ranging spectrum of environmentally friendly technologies that power the connected information infrastructure globally. The GICT Working Group (WG) within GISFI brings the stakeholder's standpoints to the standardization forum and seeks to evolve the most suitable approach for adoption in the Indian context. The WG also aims to develop specifications in specific areas where existing global standards do not meet Indian requirements. In this paper, an overview of GICT WG activities is provided along with the current progress of important work items from an Indian standards development perspective. The paper also outlines the collaboration of the GICT WG with external organizations and presents a structured roadmap of technical specifications development.

Keywords: GICT, Energy Efficiency Metrics, Telecommunications.

1 Introduction

Green Information and Communication Technology (GICT) is a Working Group (WG) of Global ICT Standardisation Forum for India (GISFI) since

Journal of ICT Standardization, Vol. 1, 205–220.
doi:10.13052/jicts2245-800X.12a6

its inception in 2009. This WG has developed three Technical Reports (TRs) [1], [2] and [3] and twoTechnical Specifications (TSs) [4] and [5]. GICT WG of GISFI has liaised with global Standards Developing Organizations (SDOs) and is working closely with organizations such as the Telecommunication Engineering Centre (TEC) under the Department of Telecommunications, Government of India.

Activities of the GICT WG of GISFI are of the utmost importance for India because ICT is one of the fastest growing sectors in the Indian economy (it contributed approximately seven percent of the Indian GDP in 2012 [8]), as well as in the world. However, the growth of ICT is a double edged sword for India: while it brings economic growth and solutions that curb the growth of Green House Gases (GHG) in other industry sectors, in itself it also contributes to pollution and impacts the environment and economy in various ways. In India, around 4% of the GHG emission is from the ICT sector which is around 80 million tonnes of CO_2 emission every year. Around 25% of this emission, i.e., 1% of the GHG emission is from the telecom sector which is around 20 million tonnes of CO_2[9].Thus there are two important aspects regarding Green ICT in India: (1). Making ICT green, and (2). Using ICT technology towards making other industry sectors green.

This paper begins with a section on the scope and objectives of the GICT WG. Thereafter, in the third section a brief outline of the achievements of GICT WG is provided. Next, in the fourth section, ongoing collaborative activities with the Government of India and Global SDOs are described. In the fifth section, the planned activities of the GICT WG in the near future are outlined. The sixth and final section concludes this paper.

2 Scope and Objectives

Green ICT applies tothe entire lifecycle of products starting from the design, to production (material and method of production), operations and goes on further to the product end of life or recycling phase. Since this spectrum was far too wide for the Green ICT WG to address effectively in the initial phase of its activity, it was decided by the WG to retain focuson (a). Greening the ICT infrastructure during its usage and (b). Identifying means to use ICT to make other industry sectors green. Thus technical and standards considerations related to the complete lifecycle of product; materials and substances used for product development and recycling were designated 'out of scope' for the GICT WG in its initial set of work items.Further, the activities of the GICT WG were guided by the technical expertise; understanding of global business implications for India; and the interpretation of unique requirements for Green ICT in India by its members, driven by a mechanism based on GISFI member consensus that represented corporates, individual contributors and academia. Significant technical contributions to the WG were received from corporate members NEC, Ericsson, Cisco, VNL, I2TB-SPPL and Graceful Growth Consulting Services; individual contributors and academic institutions like IIIT-Allahabad.

The following are the stated objectives of the GICT WG:

  i. Define metrics and methods of measurement for energy efficiency of telecommunications equipment;
 ii. Study potential enhancements of ICT for making it green; and
iii. Study potential use of ICT to make other sectors green (ex: Agriculture, Transport).

3 GICT WG Contributions

Over the last three years, the GICT WG along with stakeholders has put forth significant efforts to carry out detailed study of the standards landscape, regulations and India-specific requirements of energy efficient ICT. Figure 1 illustrates a map of significant contributions in terms of technical reports and specifications proposed by the GICT WG. In this section, an overview of activities of the WG in the following three major areas of focus has been provided:

> Subsection 3.1: *Metrics and methods of measurement for energy efficiency of telecommunication equipments*
> Subsection 3.2: *Study on potential enhancements of ICT*
> Subsection 3.3: *ICT to make other sectors green*

Complete descriptions of the contributions listed in Figure 1 are available in the documents listed in the References.

3.1 Metrics and Methods of Measurement for Energy Efficiency of Telecommunication Equipments

The WG has produced one approved technical report and two technical specifications in this area which have effectively formed the baseline for the national green telecom implementation plan of the TEC in India. The technical report [3] surveys the existing work at various international standard development organizations such as the International Telecommunication Union (ITU),

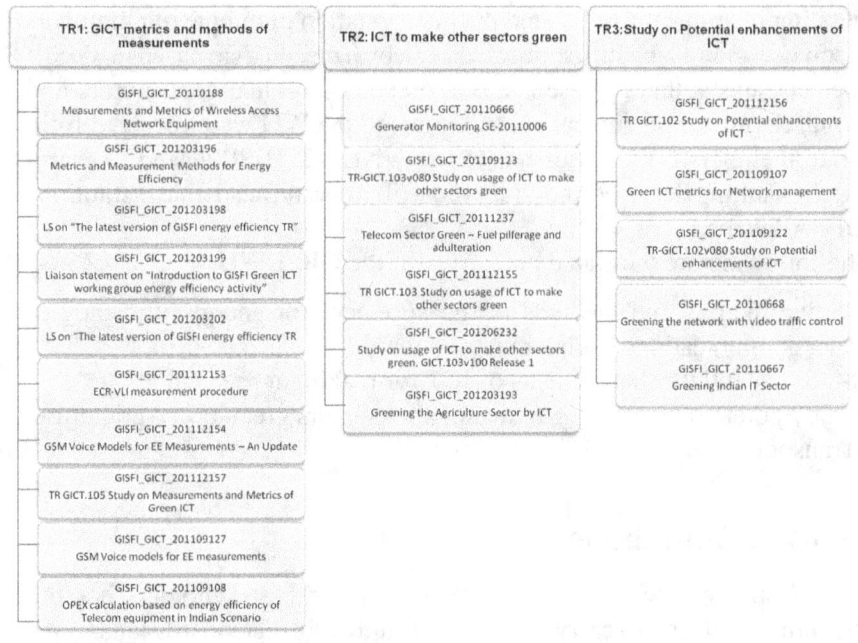

Figure 1   Standards development – GICT WG contributions map [7].

Alliance for Telecommunications Industry Solutions (ATIS) and European Telecommunications Standards Institute (ETSI). It also surveys the Telecom Regulatory Authority of India (TRAI) recommendations towards implementation of green telecommunication in India. More specifically, it describes the metrics and method of measurement for circuit switched broadband equipment (DSLAM) and GSM radio base station. The gaps in the current standards as well as TRAI recommendations are outlined towards the end of this technical report.

3.1.1 End-to-End Landscape of Global GICT Standards

Figure 2 illustrates the span of global standards from ETSI/ITU/ATIS across access network, core network and the data centres that form the complete end to end network topology of any telecom service provider.

As can be inferred from Figure 2, it is apparent that no single standard covers the entire span of the telecom network and therefore efforts such as those done by the GICT WG are needed to stitch together the range of ICT standards applicable to India. While leveraging the existing global standards,

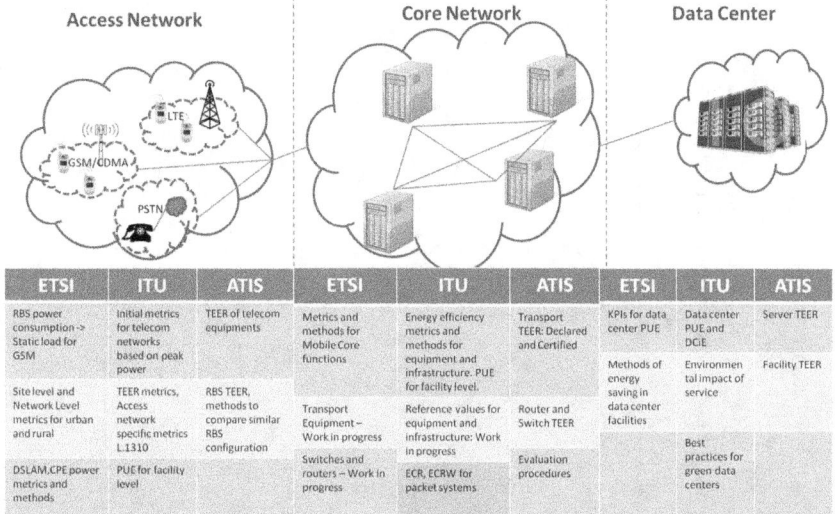

Access Network			Core Network			Data Center		
ETSI	**ITU**	**ATIS**	**ETSI**	**ITU**	**ATIS**	**ETSI**	**ITU**	**ATIS**
RBS power consumption -> Static load for GSM	Initial metrics for telecom networks based on peak power	TEER of telecom equipments	Metrics and methods for Mobile Core functions	Energy efficiency metrics and methods for equipment and infrastructure. PUE for facility level.	Transport TEER: Declared and Certified	KPIs for data center PUE	Data center PUE and DCiE	Server TEER
Site level and Network Level metrics for urban and rural	TEER metrics, Access network specific metrics L.1310	RBS TEER, methods to compare similar RBS configuration	Transport Equipment – Work in progress	Reference values for equipment and infrastructure: Work in progress	Router and Switch TEER	Methods of energy saving in data center facilities	Environmental impact of service	Facility TEER
DSLAM,CPE power metrics and methods	PUE for facility level		Switches and routers – Work in progress	ECR, ECRW for packet systems	Evaluation procedures		Best practices for green data centers	

Figure 2 Survey of existing global standards and their applicability to telecommunication equipments [7].

it is also very important to fill the gaps with new or modified standards relevant for Indian telecom networks.

3.1.2 Classification of Telecommunication Equipments

While considering the standards applicable to the end-to-end telecom infrastructure, it is also important to classify and prioritize telecommunication equipments in order to recommend or generate appropriate technical standards. In this context, the GICT WG has delivered Technical Specification (TS) [5] that detail the classification of telecommunication equipments for the purpose of energy efficiency measurement. Table 1 shows the classification model for telecommunications equipment developed by the WG under the TS.

3.1.3 Radio Base Station Energy Efficiency Metrics

If we focus on the access network side of the telecom service provider infrastructure, the Radio Base Stations (RBS) stands out as one of the most energy consuming class of telecom equipment. This is especially significant from the Indian context because of the very large installed base of RBSs (more than 400,000 sites [14]) and coupled with the acute shortage of grid power, this class of telecom equipment is considered to be of the highest priority for the energy efficiency standards activity. However,if the energy efficiency

Table 1 Broad classification of telecommunications equipment [5].

Access network	Core network	Data center

• Radio Access Network
  o Circuit Switched Base station (GSM/ CDMA, WCDMA)
  o Packet Switched Base Station (3GPP LTE/IEEE 802.16e)
 	
• Fixed Access Network
  o Circuit Switched
  o Packet Switched
  o Passive Optical Network
	• Transport Network
	  o Circuit Switched Optical Equipment
	  o Circuit Switched Non-Optical Equipment
	  o Packet Switched Optical Equipment
	  o Packet Switched Non Optical Equipment
	  o Converged Packet Optical Equipment
	
	• Mobile Core network
	  o 2G and 3G Mobile Network Equipments
	  o Packet Switched Mobile Core Network Equipments
		• Data Center equipments
		  o Servers
		  o Data Center router, switch
		  o Storage equipment
		  o Facility
		  (Data center infrastructure)

standards from ETSI/ATIS/ITU on RBS equipments are studied in detail, it is revealed that no single global standard can effectively address all Indian requirements. An overview of the comparative analysis of specifications on RBS energy efficiency metrics from various global SDOs is shown in Table 2. GISFI GICT WG is actively progressing towards framing of specifications that address all the Indian requirements related to the RBS energy efficiency metrics and methods of measurements.

The GISFI GICT Technical Specification [4] is an umbrella document that spells out the general requirements for measurement setup, procedures, reporting format and assessment scales.The GICT WG follows the ITU's general definition of energy efficiency metric for telecom equipments [11]. Further, the document also provides the generalized definition of Energy Efficiencybased on which subsequent specifications would address the requirements of each class of equipments.

> *The general definition for metric by ITU is: "Energy efficiency will be defined as the relationship between the specific functional unit for a piece of equipment (i.e., the useful work of telecommunications) and the energy consumption of that equipment." [11]*
>
> $General\ Definition\ of\ Energy\ Efficiency = \dfrac{Useful\ Work\ Done}{Power\ Consumed\ by\ Equipment}$

3.2 Study on Potential Enhancements of ICT

The GICT WG has produced one TR[1] on the study on potential enhancements of ICT sector to make it more environmentally friendly. It surveys the related topics of Green Energy and GHG Emission. Further, the TR describes the ongoing activities in ICT standardization and potential areas of improvements. Both the telecom sector as well as the broader ICT sector has been addressed in this report, and potential gaps in the current standards and scope for enhancement within the purview of GICT WG activities are addressed.The various components of ICT and their CO_2 emissions in the ICT sector and their footprints are shown in Figure 3.

GICT is the study and practice of designing, manufacturing, using and disposing of computers, servers, and associated subsystems—such as monitors, printers, storage devices, and networking and communications systems—efficiently and effectively with minimal or no impact on the environment. Thus, Green IT includes the dimensions of environmental sustainability, the economics of energy efficiency, and the total cost of ownership, which includes the cost of disposal and recycling. Figure 4 shows the percentage of benefit by practicing green IT.

Table 2 Comparative analysis of specifications on RBS energy efficiency metrics from various global SDOs.

Indian requirement	ETSI [10]	ATIS [12]	ITU [11]	GISFI [7]
Alignment to General Definition of Energy Efficiency	No	Yes	Yes	Yes
Applicability to Equipment Level Metric	No	Yes	No	Yes
Applicability to comparison amongst RBS deployment	No	Yes	No	Yes
	Only Macrocell	Both Micro and Macrocell		Both Micro and Macrocell
Support for Integrated & distributed RBS configurations	Yes	Yes	Yes	Yes
Support for variable traffic load based test	Yes. Three levels	Yes. Five levels	Yes. Three levels	Yes Three levels
Independent of UE deployment model, subscriber density information	Yes	No	Yes	Yes
Support for model of daily variation of traffic load	Yes	No	Yes	Yes

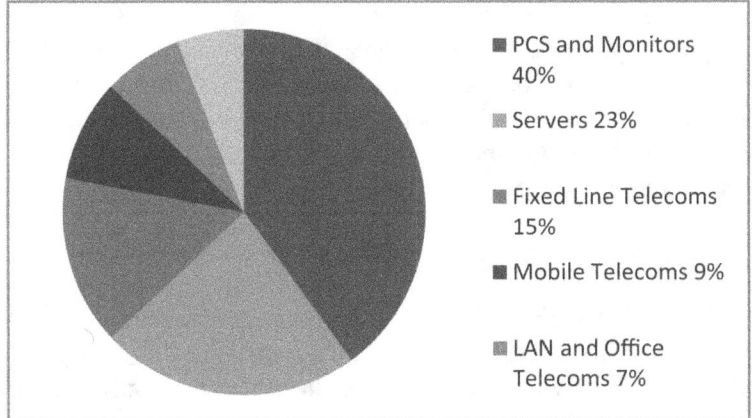

Figure 3 Distribution of CO_2 emissions in IT sector [9].

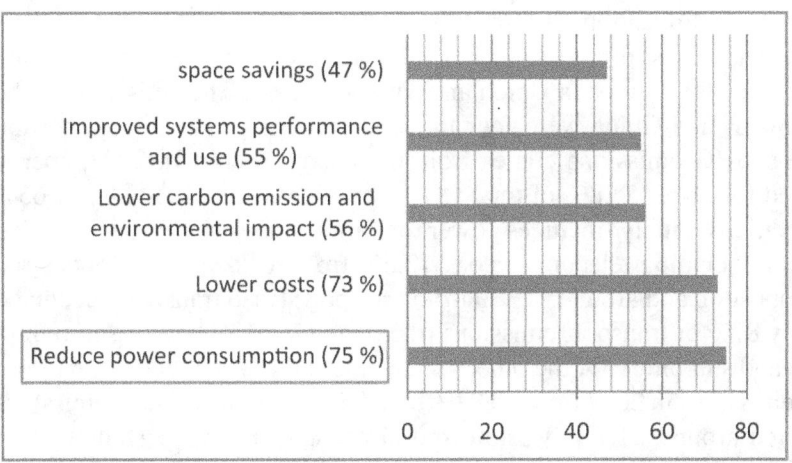

Figure 4 Green ICT adoption benefits [7].

Energy efficiency initiatives in ICT have mostly focused on consumption while in use. However, a large part of the environmental impact of a device comes during its manufacturing and disposal. The whole life impact of equipment is much more difficult to measure than the in use energy consumption, because it is spread through a long supply chain. Figure 4 depicts the benefits of adopting Green ICT technologies. It is clear that reducing power consumption of equipments is the biggest contributor to the benefits possible from Green ICT adoption hence it is the single most important factor to consider while looking to enhance the energy efficiency of telecommunications equipment. By conserving energy, GHG emissions can also be reduced significantly and also potentially reduces cost of operation for different ICT products and solutions. Understanding current ICT scenario in India enables to set the goals for reduction in carbon emissions and providing solutions that require minimum power for satisfying the user requirements.

3.3 ICT to Make Other Sectors Green

The scope of the GISFI Technical Report [2] produced on this topic is a study on the means to use ICT to make other sectors green. The TR identifies ongoing standardization activities worldwide on the topic of using ICT to make other sectors green. Further Indian scenarios are studied and requirements on using ICT to make other sectors green are identified. Sectors within the scope of the technical report are: Power sector, Transportation, Manufacturing, Agriculture, E- Education, Data-centres.

India was ranked fifth globally in total GHG emissions, behind the United States, China, the European Union and Russia in 2007. The emissions of the United States and China were four times that of India. The largest portion of India's GHG emissions comes from the energy sector. In 2007, energy accounted for about 57.8% of total CO_2e emissions – of which almost 65% from electricity supply (includes power for various industries through captive generation including telecom sector), 12.6% from residential and industrial fuel combustion and around 12.9% from transport. Road transport accounted for nearly 87% of transport emissions (the remaining 13% coming from rail, aviation and shipping). Of the other sectors, agriculture accounted for 17.6% of total emissions in 2007 (around 22% in 2005), industrial process emissions contributed around 21.7%, waste disposal accounted for 3% (falling from nearly 7% in 2005), and land use and land use change (LULUCF) accounted for 9% (net carbon storage in 2007). Figure 5 shows a sector-wise breakdown of emissions for 2007 [13].

There are opportunities for reducing emissions through intensive use of ICT across many sectors like energy creation and distribution, buildings, transport and industry. Preparing core ICT infrastructure for these technologies is critical. ICT can be used for disaster relief/early warning and for emergency services, which are particularly important in mitigating the effects of climate change, for instance from flooding or increased incidence of violent storms and hurricanes.In the initial phase, GISFI GICT WG has been working in the following sectors to reduce GHG emissions:

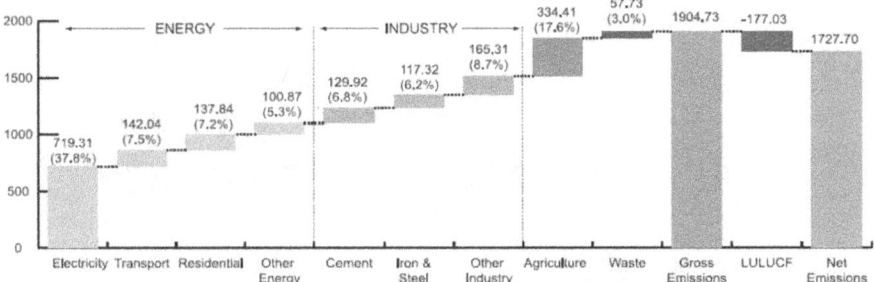

Figure 5 Sector wise GHG emissions in 2007 (million tons of CO_2 eq) [13].

- *Telecom Sector*
- *Energy Sector*
- *Transport Sector*
- *Agriculture Sector*
- *e-Education Sector*

4 Collaboration with Government and Other SDO's

The GICT WG has forged strong working relationship with Government of India organizations like the TEC, Department of Telecommunications. GISFI members actively participate in the core committee setup by the TEC towards formulating the guidelines for the implementation of Green Passport certification for telecommunication equipments in India. The Figure 6 shows the most important outputs from the GICT WG. In the period from 2011 to 2013, the WG has released three TRs and two TSs on topics described in the Section 3. It may be noted that the GICT standardization is an ongoing process in which stakeholders comprising industry, government and academia are closely working with each other to develop specifications and standards.

The GICT WG has also sent liaison statements to the ITU-T Study Group 5 as well as other SDOs such as ATIS to collaborate closely on measurement metrics and methods of energy efficiency for telecom equipments. Global SDO representatives like the ITU-T SG5 have participated in the GISFI Standardisation Series Meetings in the past and have committed their support for GISFI activities.

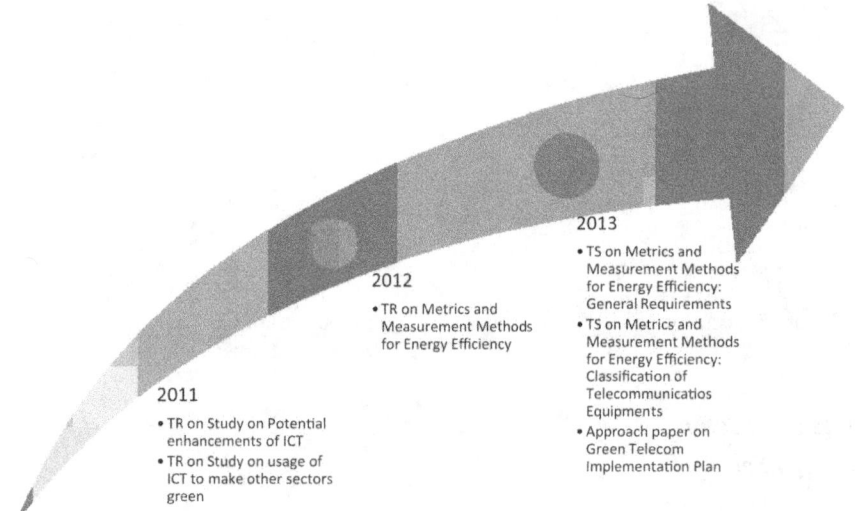

Figure 6 Key contributions from the GICT WG [7].

5 Future Plans

The GICT WG has a well-defined roadmap for specifications and standards development. Figure 7 shows the GISFI GICT WG Standards Roadmap for the creation of technical specifications in the focus area of telecom equipment energy efficiency. In addition, the GICT WG in partnership with government organizations such as the TEC, would help spearhead the green telecom standardisation and certification efforts in India.

Development of the following GISFI Technical Specifications (Table 3 and Table 4) are forthcoming in the near term based on the urgency and prioritization of efforts determined by stakeholders across the Government of India, Industry and Academia:

6 Conclusions

This paper reviewed the importance of developing Green ICT Standards for India, and provided an overview of efforts of GISFI by citing Technical Specifications and Technical Reports developed by the Green ICT Working Group. The longer-term work plan of the GICT WG was elucidated via a Standards Development Roadmap that was constructed in consultation with its stakeholders. GISFI GICT Working Group continues to work intensively

TS0: Metrics and Measurement Methods for Energy Efficiency: General Requirements

1st version completed and Approved by GISFI

TS 1 :Metrics and Measurement Methods for Energy Efficiency: Radio Access Network

RELEASE 1	RELEASE 2
Circuit switched radio base station	Packet switched radio base station
a. GSM , CDMA – Narrowband 2G	a. IEEE 802.16e (WiMAX BS)
b. WCDMA – Wideband 3G	b. 3GPP LTE (LTE eNodeB)

TS2: Metrics and Measurement Methods for Energy Efficiency: Fixed Access Network

RELEASE 1	RELEASE 2
Circuit switched equipment	Packet switched equipment
a. Broadband end user equipment (Customer Premise Equipment)	a. Broadband end user equipment (Customer Premise Equipment)
b. Digital Subscriber Line Access Multiplexer (DSLAM) Equipment	b. Access Ethernet router with Ethernet on LAN side
c. Access DSL router with Ethernet on LAN side	c. Access Ethernet router with wireless function on LAN side
d. Access DSL router with wireless function on LAN side	d. Access Ethernet high speed switch (L2 switch)

TS3: Metrics and Measurement Methods for Energy Efficiency: Transport Network

RELEASE 1	RELEASE 2
Circuit switched optical equipment	Circuit switched equipment – other than optical
a. Multiservice SONET/SDH ADMs, MSPP, optical cross connect (OXC) systems, digital cross connect systems (DCS)	a. Free space optics
b. Optical Transport network (OTN) equipments	b. Point to Point wireless transport
c. ROADM/WDM and similar equipment	
	Packet switched equipment – other than optical
Converged packet optical equipment	a. Point to Point wireless transport
a. Independent switching of TDM signals and packet signals in both directions	b. Distribution or Aggregation L2 Switch
	c. Core L2 switch
b. Partial conversion from TDM to packet signals and vice –versa prior to switching	d. Edge Router
	e. Core router
c. Full conversion from TDM to packet signals and switched via the packet switch	**RELEASE 3**
	v. Specialized transport equipment
d. Full conversion from packet signals to TDM and switched via the TDM switch	a. Video transport equipment
	b. Storage area networking equipment

TS4: Metrics and Measurement Methods for Energy Efficiency: Mobile core network

RELEASE 1	RELEASE 2
2G and 3G mobile network	Packet switched mobile network
a. GSM, UMTS core network equipment	a. IMS core network equipment
b. GSM, UMTS radio access control network	b. EPS core network equipment

TS5: Metrics and Measurement Methods for Energy Efficiency: Data centres

RELEASE 1	RELEASE 2
a. Servers	a. Facility level metrics
b. Data center router	
c. Storage equipment	

Figure 7 GICT WG standards roadmap [7].

Table 3 TS on metrics and measurement methods for energy efficiency: radio access network.

Release 1	*Release 2*
Circuit switched network radio base station	Packet switched network radio base station
a. GSM, CDMA – Narrowband 2G	a. IEEE 802.16e (WiMAX BS)
b. WCDMA – Wideband 3G	b. 3GPP LTE (LTE eNodeB)

Table 4 TS on metrics and measurement methods for energy efficiency: ip router and switch.

Release 1	*Release 2*
a. Access and Edge Router	a. Core router
b. Access and Distribution Switch	b. Core and data center switch

with stakeholders to achieve the goals illustrated in the roadmap. Green ICT is an area of great importance and focus for the Nation of India, the Industry, Research Fraternity, Policy Makers and Academia alike. The efforts of GISFI and those of the GICT WG are aligned to the overarching goal of creating a green and sustainable planet and delivering the promise of ICT technologies through standards development that will drive global adoption.

References

[1] GISFI TR GICT.102 V1.0.0, Study on Potential enhancements of ICT; Release 1, December 2011.

[2] GISFI TR GICT.103. V1.0.0, Study on usage of ICT to make other sectors green; Release 1, September 2011.

[3] GISFI TR GICT.105 V1.1.0, Metrics and Measurement Methods for Energy Efficiency; Release 1, December 2012.

[4] GISFI TS GICT.100 V1.0.0, Metrics and Measurement Methods for Energy Efficiency: General Requirements; Release 1, January 2013

[5] GISFI TS GICT.101 V1.0.0, Metrics and Measurement Methods forEnergy Efficiency: Classification of Telecommunication Equipments; Release 1, February 2013

[6] Approach towards Implementation of Green Telecom in India, GISFI GICT WG, January 2013.

[7] GISFI Meeting Documents: Green ICT URL: http://gisfi.org/workinggroups.php?wg= GICT [Last accessed: 13th July, 2013]

[8] Indian IT-BPO Industry, NASSCOM URL: http://www.nasscom.in/indian-itbpo-industry [Last accessed: 13th July, 2013]

[9] Recommendations on Approach towards Green Telecommunications, TRAI URL: http://trai.gov.in/Content/RecommendationDescription.aspx?RECOMEND_ID=232& qid=0 [Last accessed: 13th July, 2013]

[10] ETSI TS 102 706 V1.2.1; (10/2011); Environmental Engineering (EE); Measurement Method for Energy Efficiency of Wireless Access Network Equipment.

[11] ITU recommendation: L.1310; (11/2012); Energy efficiency metrics and measurement for telecommunication equipment.

[12] ATIS-0600015.06.2011; (11/2011): Energy efficiency for telecommunication equipment: Methodology for measurement and reporting of radio base station metrics.

[13] Ministry of Environment and Forests, Government of India, India: Greenhouse Gas Emissions 2007 URL: http://moef.nic.in/downloads/public-information/Report_INCCA.pdf [Last accessed: 13th July, 2013]

[14] Powering Cellular Base Stations: A Quantitative Analysis of Energy Options- Prof. Ashok Jhunjhunwala, RITCOE- IIT Madras Solar PV, Diesel Generators, Batteries and

Electrical Grid URL: http://www.tcoe.in/download/1/Download-Section_190/1/1.html
[Last accessed: 13th July, 2013]

Biographies

Ritesh Kumar Kalle is Research Engineer at NEC Mobile network Excellence Center (NMEC), Chennai, India. Prior to this, he was the Senior Scientific Officer and a full time PhD scholar in Wireless Networks Lab at IIIT-Bangalore. He primarily works in related areas of modelling and performance analysis of Mobility, Power Saving and Quality of Service in Wireless networks, particularly LTE, WiMAX and Wi-Fi. His research was supported by Microsoft Corporation under the Microsoft Research India PhD Fellowship Award and the Departmentof Information Technology, Govt. of India. Ritesh has a graduate degree in Information Technology from IIIT-Bangalore. He has applied several patents and has published extensively in peer reviewed International Journals and Conferences.

Mr. Arvind Mathur is Strategic Technology Officer, India & South Asia with Cisco's Corporate Research & Advanced Development Group and is based in Bangalore, India. Arvind's professional career spans over 23 years in the telecom & ICT industry and he brings extensive service provider executive leadership experience from India as well as internationally. Prior to joining Cisco in 2010, Arvind was CTO and President, Global Services at Sify Technologies; CTO at Bharti Airtel for Enterprise Services; and Vice President & Global Head of Product Management for Enterprise Network Services at Tata Communications, formerly Tata-VSNL.

Arvind has spent several years in Japan, USA and Canada working in different roles and capacities with the Research Institute of Electrical Communications, Teleglobe International, and JDS Uniphase. At Cisco, he is working on several initiatives for Service Providers, Public Sector and Government Affairs, contributing to thought leadership in technology, service architectures, solutions, standards and policy. Arvind is a familiar industry speaker on next generation converged ICT networks, Internet of Everything/Internet of Things, cloud computing, data centers, managed services and innovation strategies. Arvind is an alumnus of the Indian Institute of Science, Bangalore and the Indian Institute of Technology, Delhi.

Carbon Emission Estimation & Reduction in Indian Telecom Operations: A Standardization & Policy Perspective

Krishna Sirohi

President – Impact Innovations in Technology & Business (i2TB), India;
e-mail: president@i2tb.in

Received July 2013; Accepted August 2013

Abstract

This Paper is Standardization Perspective of the problem definition of Carbon Emission Estimation and Reduction exercise in Indian Telecom Networks. It includes the relevant interpretation of the TRAI recommendations, current practices of estimating Carbon emission and the associated issues. It brings out the techno-managerial perspective to estimate carbon emission in Indian Telecom network, description of with the currently adopted methods and their implications. It also suggests how the entire telecom sector may be geared up to identify all potential areas for energy efficiency optimization in every part of the telecom network such that all parts of the telecom networks are addressed to achieve significant reduction in Carbon Emission. Some key aspects related to necessary framework for accurate estimation, setting realistic national targets for carbon emission reduction, automatic online gathering of carbon emission parameters from Infrastructure provider (IP) and telecom equipment based on IoT Framework have been presented for suitable consideration for related policy formulation. The paper summarizes the need for Standardization and policy formulations actions to achieve good progress in making telecom network infrastructure sustainable.

Journal of ICT Standardization, Vol. 1, 221–240.
doi:10.13052/jicts2245-800X.12a7

Keywords: Carbon Emission Estimation & Reduction, Carbon Foot-printing, TRAI, Traffic Profile, Green Telecom, Telecom Service Provider (TSP), Infrastructure Provider (IP), Renewable Energy Technology (RET).

1 Introduction

India, as a globally responsible nation, has accorded high importance and committed to reduce carbon emission from its various sources by a definite percentage in a time-bound manner. The percentage contribution of the Telecom Network with respect to other carbon emitting sectors is insignificant currently, but considering the expected growth in the telecom networks especially in rural area with poor grid power situation and with the increased data services, its potential for a significant rise is very high. The national leadership (regulatory and telecom service providers) is determined to achieve lowest level of carbon emission by leveraging all opportunities ie, developing new and adopting energy efficient (green) telecom technology in Indian Network, optimizing all operational aspects leading to a minimum energy consumption and by adopting Green Energy sources to run the telecom infrastructure. Timely initiative of Indian Telecom Regulatory Authority (TRAI) of releasing its "Recommendations on Green Telecom" in consolidating entire telecom sector stakeholders opinion and providing clear direction for achieving green telecom objectives and later transforming them into mandatory operational requirements to Telecom Service Providers (TSP) is a strong step in the direction.

Currently, a very large proportion of the total telecom networks carbon is emitted by Diesel operated Generators to feed power to mobile network Base Station Sites. Passive infrastructure of these base station sites that includes power, tower physical space and security/operating personals of the telecom networks is being managed by non-Telecom Service Providers (TSP) business entities called Infrastructure Providers (IP). IP offers these services on shared basis for many TSPs as the tenants of their facility.

According to the prevailing Green Telecom Regulations, TSPs have started working out total Carbon Emission Estimates (CEE) of their respective Telecom Network, started submitting to DoT and TRAI on half yearly basis where this data is expected to be analyzed for to accuracy verification and to assess the expected mandatory reduction in CEE on year by year basis.

TSPs are dependent on Infrastructure Providers (IP) for providing major accurate data required for calculating the carbon emission for their base-station sites. The data being providing by Infrastructure Provider (IP) is normally questioned for its accuracy. The problem is further aggravated when the

CE iscalculated based on the formula recommended by TRAI that accounts Carbon emission based upon the Maximum power generating capacity of the Diesel Power Generator used to power large number of Basestation sites, not on the basis of the power consumption of the Telecom Equipment (BTS with is associated cooling system). Boththe factors collectively contribute to a significantly higher estimate of carbon Emission than actual. The extent of error is estimated to be of the order of the reduction targets set for achieving during the next few years. The current situation where Diesel Power Generators are considered to be dirty source of energy for meeting the power consumption requirements to the Mobile Radio Basestation is accorded the highest importance. The entire focus of the entire Telecom Sector in India is only towards replacing Diesel Power Generation sites by Renewable Energy Technology (RET). All the techno-commercial feasibility issues related to Diesel Generator substitution with RETs are being treated as if it is the only cause for poor carbon emission state of Indian Telecom. This perception has created insensitivity towards achieving Energy Efficiency in other parts of the segments of the Telecom Networks. Adequate timely attention to achieve energy efficiency in the telecom network with evolution of necessary Standards for measurements and metrics of energy efficiency, development of energy efficient technology building blocks, telecom products and the operating methods must be treated with equal importance.

Additionally, there is an urgent need,in the country,for necessary framework of the consolidated data used for estimation of Carbon Emission from all the TSPs as national Telecom Carbon Emission Platform. The platform is to be utilized by the Government and the Industry to draw useful intelligence as required to formulate a realistic national plan for Carbon Emission Estimation & Reduction (CEER) and also provide the firm direction to Telecom Service Providers, Telecom Equipment manufacturers and the Telecom Technology innovators for achieving national goals of higher energy efficiency in Indian Telecom and realize a sustainable telecom infrastructure in India. The same can be achieved by evolving and adoptinga holistic approach at a national level so that all the parts of the telecom networks be subjected to separate energy efficiency review and optimization by adopting currently available energy efficient Telecom Equipments, by motivating and supporting development of innovative energy efficiency technology of Telecom equipment, optimizing operating methods for achieving energy efficiency in telecom operation and also by converting polluting to greener energy sources.

This paper discusses current methods, associated issues and proposes some perspective that would be necessary for evolving necessary holistic plan for

Carbon Emission Estimation and Reduction for India's Telecom services. This approach will set a technically sound ground for setting up reasonable expectations and deterministic approach of achieving the set national targets for telecom carbon emission. Detailed and holistic approach as proposed in the paper helps in identifying all potential problematic areas to draw equal attention from Government and Industry for its resolution. The expected outcome include the identification of all areas of energy consumption sub-optimality, definition of required policy/methods for addressing them leading to achieve set objectives and to establish India's global leadership in the field of telecom network carbon emission reduction.

2 Parts of Telecom Networks and Their Respective Carbon Emission Estimates

Referring TRAI Recommendation [1], one of the possible classifications of entire Telecom Networks into various network segments has been presented in the following table (Table 1):

The average percentage proportion of the power consumption and hence estimated carbon emission from these network segments have been presented in Figure 1 above.Closer view of the Figure 1 suggests that the major most portion of the carbon emission is contributed by the Mobile network. The base stations alone contribute to 59% of the total carbon emission of the entire mobile network. Remaining 41% carbon emission of the entire mobile network is being contributed by rest of the network. The carbon emission estimates of various segments of the telecom network sets the priority for addressing carbon emission problem.

3 Problem Defintion of Carbon Emission Estimation (CEE)

The carbon emissions estimation depends on two factors i.e. the power consumption of the telecom equipment (along with its mandatory cooling system requirements) and the carbon emission factor associated with the power source being used for powering the telecom equipment.

The first part of the problem is the right measurement of the power consumption of any telecom equipment deployed in the live network that variessignificantly over the time cycle of 24 hours. There must be a appropriate method to measure accurate power consumption that is required for the

Table 1 Telecom network classification.

S.No.	Network Segment	Comment
1.	Access Network	
a.	Landline (C_L)	Carbon Footprint of Landline Access Network
i.	Exchanges-Local, Tandem, TAX (C_{Le})	
ii.	Copper Distribution Network (C_{Lc})	
iii.	Telephones (C_{Lt})	
b.	Mobile Access network (C_M)	$C_M = C_{MS} + C_{BTS}$
i.	Main switching Centers (C_{MS})	
	MSC	
	Media GW	
	SGSN	
	GGSN	
	BSC	
ii.	Base Station Controller Centers (C_{BSC})	BSC is include in Main Switching
iii.	Base Transceiver Station (C_{BTS})	This is entire BTS Site power requirement including BTS Equipment, BTS Side Transmission equipment and Cooling System
iv.	Mobile Phones (C_{BS})	Not estimated currently
c.	Fixed Broadband (C_{FB})	
d.	Fibre (FTTx) (C_F)	

(Continued)

Table 1 Continued

S.No.	Network Segment	Comment
2.	Core Network (C_C)	Applicable for all Unified Access Service licensee
a.	IP-Cores	
	Edge / Core Routers	
	NGN Nodes/ IMS Nodes	
	Soft-switches	
	Data-Centers	
	HLR	
	SCP (IN)	
	SMSC	
	OSS/BSS	
	VAS	
	All other centralized sub systems / peripherals	
3.	Aggregator or Backhaul (C_A)	Transmission equipment at Main Switching equipment side. BTS side of Transmission Equipment is included in C_{BTS}.
4.	Transmission Network (C_{TX})	
5.	Infrastructure Provider (C_{IP})	This include the entire Carbon footprint of Basestation where IP provides passive infrastructure. BTS sites managed by TSP is included in C_{BTS}

Total Carbon Footprint $(C_T) = C_L + C_M + C_{FB} + C_{FT} + C_C + C_A + C_{TX} + C_{IP}$

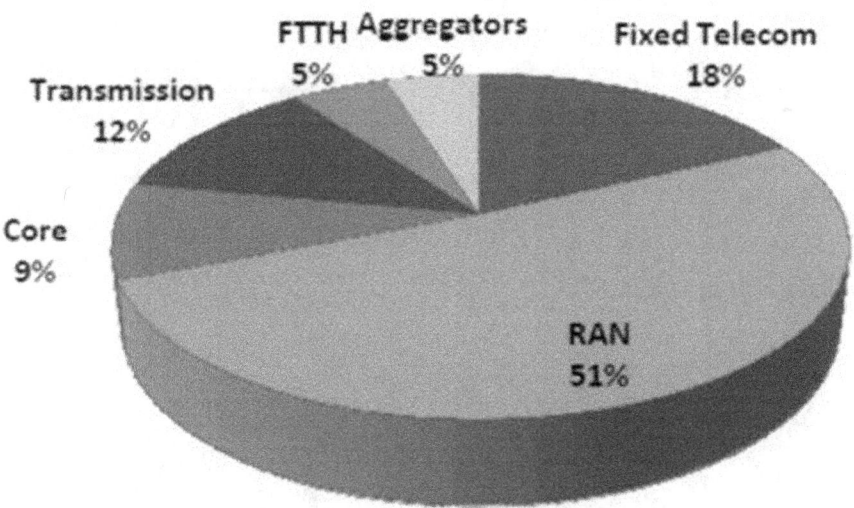

Figure 1 Power consumption breakup.

estimation of Carbon Emission. This problem will be dealt in more details in later section.

The second part of the problem i.e. the carbon emission association with the power sources used to power the telecom equipment power need. This factor is particularly prevailing in the developing world where the Grid power is not sufficient to run the telecom infrastructure. Availability and the quality of Grid power in various parts of the country varies significantly. Particularly in the rural areas, where grid power availability is extremely poor, the telecom services majorly depends upon the power generated by burning Diesel as fuel in large number of power generators deployed across the country.

In general, any Telecom Equipment's power consumption requirement P can be considered as being met by various power sources i.e. Grid, Diesel or any Renewable Energy Technology (RET) like Solar. For each type of power source there is an associated Carbon Emission Factor i.e. CF_g for Grid Power, CF_d for Diesel and CF_r for Renewable Energy Source. The problem definition is presented in Figure 2.

If the power consumption need of the telecom equipment is met for T_g Duration from Grid, T_d from Diesel and T_r from Renewable Power Source

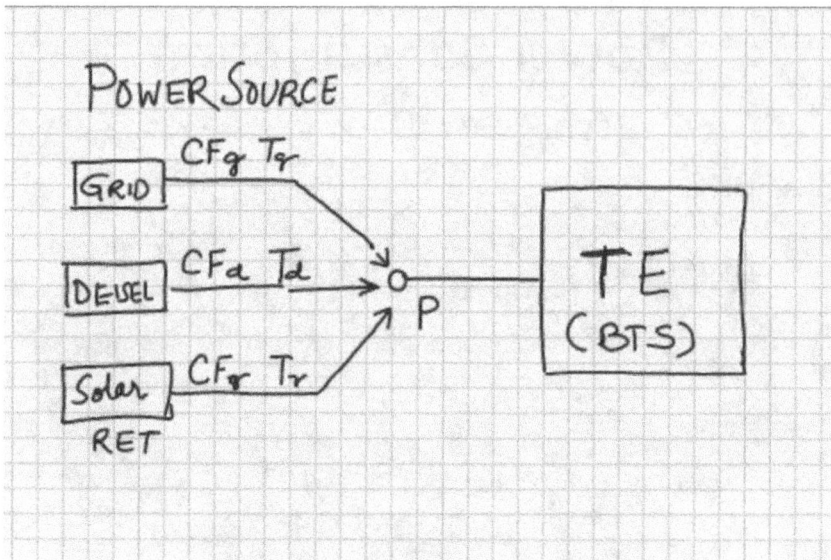

Figure 2 Carbon emission problem definition.

during a day cycle (24Hrs), the carbon emission of the telecom equipment during that day can be viewed as:

$$\text{Carbon Emission CE} = P \times CF_g \times T_g + P \times CF_d \times T_d + P \times CF_r \times T_r$$

Considering the significant variation in different Carbon Emission Factor (CF) and Duration (T) for a given telecom equipment, the above formula motivates one to:

i. Reduce the power consumption of the telecom equipment itself, and
ii. Reduce the duration of the power source which has high Carbon emission Factor (CF).

First part of the problem is to reduce the Power consumption of the telecom equipment is the mandate to Technology Researchers and for the innovators in Telecom technology and Equipment design. The size alone of the total telecom infrastructure requirement in India necessitates high focus to be accorded by policy makers and the telecom technology/product developer and innovators.

Second part of the problem is to reduce the duration of the Carbon-Expensive power source (like Diesel burning) or eliminate it altogether wherever technologically and economically feasible.

Both the problems needs different set of policies, investments, and expertise to solve. Therefore, both the problems need timely, separate and adequate attention from the concerned people or authorities, so that the carbon emission problem can be rightly understood and addressed completely. Attempt to resolve just one of the problems, as it appears to be the current case in India, will lead to achieve limited results. This paper suggests necessary timely action for the first part of the problem.

4 Dynamic Power Consumption Based Accurate Estimation of Carbon Emission

Another aspect for considerations is the true nature of the power consumption for any telecom equipment. Considering the 24 hours usage profile of any telecom equipment, the power consumption is never static. It changes significantly depending upon the traffic that the telecom equipment handles at any point of time. The traffic experienced by the telecom equipment varies with the time of the day based on various traffic generating characteristics

It is important to note that the *average dynamic power consumption of any telecom equipment is significantly lower than the static maximum power consumption*. The Carbon Emission Estimates as currently done based on the maximum power consumption of the telecom equipment, are much higher than the actual estimates. This error in carbon-footprint estimates, alone, may come out to be more than the carbon emission reduction targets that has been set to be achieved during the next five years. Therefore, it is necessary to establish the total CEE baseline (as decided to be 2011 year) for any TSP accurately and should be based on dynamic power consumption.

The important fact of estimating carbon emission based on dynamic power consumption was well recognized by Telecom Standards Development Organizations like ETSI[4][5], ATIS[6], ITU-T & GISF[2][3]I. All graduated by evolving their energy consumption measurement methods based on the Dynamic Power consumption. The external organizations ETSI, for example, evolved a high level framework for developing the energy efficient measurement standards based on dynamic power consumption for the Mobile Networks based on 3G & LTE wireless technologies. They did not find it relevant to do the same for the GSM network considering the limited life of the GSM network in Europe. However considering the high relevance in India, GISFI decided and achieved progressed to develop the same for GSM network.

The Energy Efficiency Standards as developed by the Global Standards development organization are high level frameworks with necessary guideline to be adopted.This necessitates the detailed exercise to be carried out by the Telecom Service Providers themselves to evolve necessary parameters based on the experienced traffic in their respective networks. Additionally, the traffic models that are considered in developing such frameworks by global standards development organization significantly varies from the one experienced in India and particularly the mobile networks deployed in Rural India. *India's high population density and the demography and the extent of the available telecom service in Rural areas have created absolutely new traffic models that need to be studied by the local organizations and the existing gaps in the current global standards must be filled.* GISFI is making good progress in its chosen mandate of such study and developing required national standards. Telecom Engineering Center (TEC) of Government of India, in its currently ongoing "Green Passport" definition exercise has established such India Specific needs that must be filled by Indian initiative.

This exercise will include any active traffic analysis to define the relevant (say k different types) Traffic classes (TC), prevalent duration (T_k) of each traffic class and the measurement of dynamic power consumption P_k for each traffic class for all the network elements deployed in the telecom network. The concepts is summarized in Figure 3.

In this scenario, the total power consumption of the telecom equipment during the day (24 hrs) can be calculated as:

$$P_{dynamic} = \Sigma P_k \times T_k$$

for all k traffic Classes

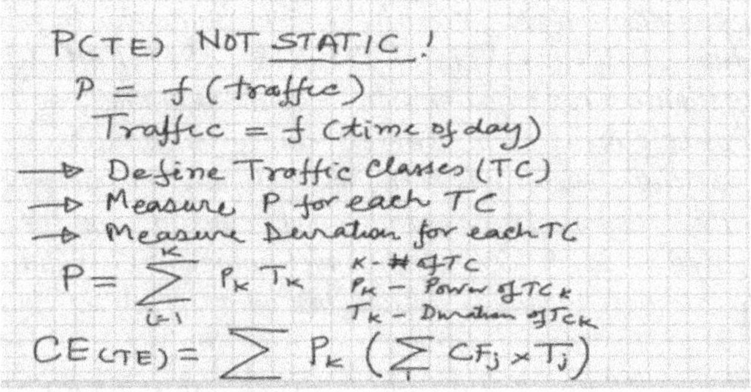

Figure 3 Carbon emission based on dynamic power consumption.

And therefore Carbon Emission for the telecom equipment can be calculated as:

$$CE_{dynamic} = \Sigma P_k \times (\Sigma CF_j \times T_j)$$

for all 'j' types of different power source with their different

applicable Power Emission Factor.

5 Benefits of Dynamic Power Consumption Based CEE

Implementation of dynamic power consumption based Carbon Emission Estimation methods has potential to lead to following *significant advantages*:

i. CEE is *accurate and significantly less than the current estimates*. This will improve Indian telecom benchmark of carbon emission on per subscriber basis in comparison of international figures.

ii. The traffic analysis as required to carry out this dynamic power consumption assessment, will lead to *identify various existing potential cases of sub-optimal energy utilization*.

iii. Identification of prevailed such cases of sub-optimal energy efficiency will motivate Telecom Service provider's operation experts to align telecom equipment design experts to evolve necessary design refinements to eliminate the sub-optimality. This will lead to a significant potential CEE reduction and hence *significant power saving i.e. the OPEX reduction*.

6 Current Methods for CEE and the Associated Issues

As discussed in the earlier sections (3), the carbon emission of a telecom equipment which power consumption needs are met by multiple power sources with its associated carbon emission factor and the duration of each power sources as application is best represented as:

Carbon Emission $CE = P \times CF_g \times T_g + P \times CF_d \times T_d + P \times CF_r \times T_r$

One this generic situation is reduced to a case when the power needs are met by two sources i.e. Grid and Diesel, the above formula should get reduced as:

Carbon Emission $CE = P \times CF_g \times T_g + P \times CF_d \times T_d$

According to the TRAI recommendations, the above formula is applicable even for the cases when the power needs are met by power source based on

Renewable Energy Technology (RET), because the carbon emission factor due to RET based power sources are considered as Zero and these power sources are considered as perfect Green Energy Sources. Additionally, TRAI has not considered the carbon emission for the duration when the Telecom equipment is run from battery and the battery is charged by Non-Green Power.

Significant concerns is raised by the Industry on the formula that has been recommended by TRAI for calculating Carbon Emission when the Telecom Equipment power requirement is met by Grid as well as alternate power source based on Diesel Generator. TRAI's recommendedformula can be viewed as:

$$\text{Carbon Emission CE (TRAI)}$$
$$= P \times CF_g \times T_g + DG \text{ Capacity} \times CF_d \times T_d$$

According to the above formula, there are two parts of the carbon emission. The first part is when the Telecom Equipment power requirement is met by the grid power and the part of the carbon emission depends on the power consumption requirement of the telecom equipment. It is to be noted that the power consumption requirement includes the Telecom equipment power consumption as well as the power required for the equipment mandated cooling system. The second part corresponds to the duration when the Telecom equipment power needs are met by Diesel Generator (DG). In this part, the power consumption requirement of the Telecom Equipment becomes immaterial and is substituted by the Power Generation capacity of the DG. The Power Generating Capacity of the DG is always higher that the power requirement of the telecom equipment. The simple reasons for putting up higher capacity DG are two. The power requirement of the telecom equipment may increase gradually in coming years due to higher demand of the traffic and there is an efficiency loss of the DG, due to aging or operational issues.

When TRAI expects TSP to calculate the Carbon Emission as per their recommended formula, one should notice that the power consumption requirement of the telecom equipment has become absolutely immaterial. This phenomenon when clubbed with large number of rural sites where Diesel consumption is high, becomes so prominent that the energy efficiency aspect of the telecom networks practically vanishes.

The DG capacity deployed in the field is not just slightly higher to cater to the factor mentioned above, but it is practically much higher. Typical DG capacity for non-shared site is 15KVA and for shared site, it is minimum 25KVA, while the power consumption requirement is typically 3KVA for non shared, 10KVA for shared sites. This is one major cause for escalated estimation of Carbon Emission in Indian Telecom Service provisioning. Situation is

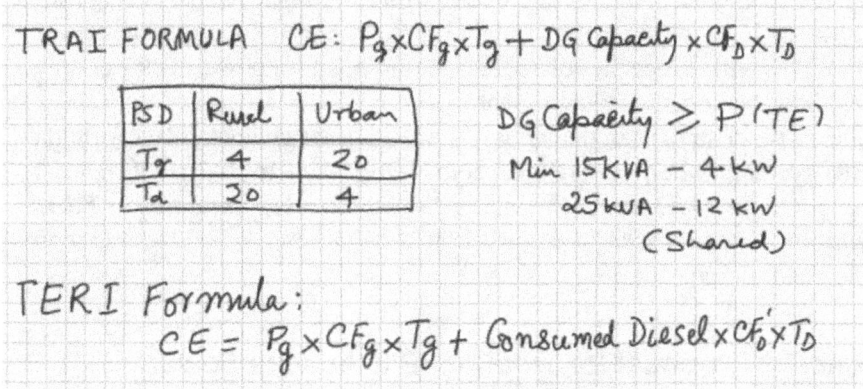

Figure 4 Diesel dominating access carbon emission estimates.

described in Figure 4 above. *The access power generator at the Basestation sites by the Infrastructure providers (IP) is the increased revenue to the IP, increased (wasteful) operating expense to TSPs which are being billed for the access power and unreasonably high carbon emission from Indian telecom networks degrading environment.* TSPs are sufferer of the increasing power expense and unfortunately have no control to reduce it.

The problem becomes more critical when the typical duration for active DG operation become as high as 20 Hrs (out of total 24 Hrs in a day) in remote rural areas.

The accumulated effect of these two factors has contributed to the TRAI's alarming calculation of 21kg of carbon emission per subscriber in India against the international figure of 8kg/per subscriber.

With these factors, the entire Carbon Emission Estimation and Reduction(CEER) exercise in India is reduced to the single agenda of converting Diesel power source to Renewable Energy Technology based power sources. This exercise requires huge investments and TSP are finding it extremely difficult to justify the investment if it is to be executed from their internal accruals. *More Governmental support is demanded, if the TSPs are expected to meet the current Green telecom mandatory regulations.* The extent of the DG-RET power source conversion problem, especially with current deployment of *energy inefficient mobile network base-station* telecom equipment is assessed so huge that there is no practical solution visible in near future.

The problem will take different shape and size and possibly get changed from the currently 'unviable' to 'viable' category, if the DG-RET power source conversion problem is judiciously clubbed with adoption of higher energy efficient base-station equipment with suitable deployment architectures, wherever possible. This will reduce the solarpanels requirement for powering the energy efficient base-station sites with a great factor.

Telecom industry often demands TRAI to atleast change TRAI formula to TERI formula where the DG capacity is replaced by the actual Diesel consumption. *TERI formula also does not address the key limitation of TRAI formula and is equally sub-optimal;* however it shows 8-10% less carbon emission estimation.

7 Concept Proposal for Treating IP'S Access Power Generation

TRAI's view of accounting entire carbon emission due to Diesel generator at the Basestation site sounds logical. The logic is based on the fact that the sole purpose to run the diesel generator at Basestation site is only to provide power to the BTS equipment. Otherwise, there was no other reason for running the DG at Basestation site location leading to any carbon emission.

However, it is meaningful to have comprehensive view of Carbon Emission Estimation problem definition. It calls for consideration of Infrastructure Provider (IP) as a business entity that can play increased role, accountability and increased business opportunity in the overall exercise of Carbon Emission Estimation and Reduction for TSPs. The IP can be visualized a responsible business entity that provide Power from multiple power sources like Grid, Diesel or RET (Solar) and meeting the need of the multiple TSPs as per the actual need of the Telecom Equipment.

In the process, IP happens to generate some access power that has a potential to be utilized for non-telecom purposes. Irrespective of the proper utilization of the access non-telecom power, it is meaningful to account this access power as non-telecom power generated by IP. Therefore total carbon emission of an IP site (shared among TSPs) can be accounted as two factors, one that is due to TSP actual power consumption requirement and another due to access non-telecom power. The same concept has been summarized in Figure 5.

Carbon Emission of an IP (Infrastructure Provider) as defined in TRAI recommendation may suitably be changed as

$$CE_{IP} = CE_{NON\text{-}TELECOM} + \Sigma CE(TSP_i) \text{ for all sharing TSPs}$$

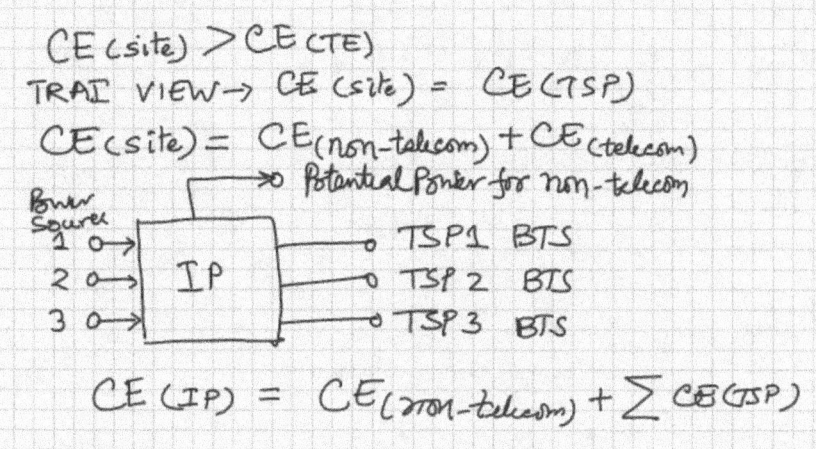

Figure 5 Non-telecom carbon emission.

Carbon Emission for all sharing TSP must be based on the actual power consumption ratio. Till the time the access power at IP remain unutilized, the same Carbon Emission for the non-telecom could also be distributed to TSP in the ratio of their power loads, but should remain separately accounted.

Similarly the total carbon emission for any TSP should include a separate entity as $CE_{NON-TELECOM}$ that lead the currently specified TRAI formula for total Carbon Emission calculation for Telecom Service Provider as:

$$C_T = C_L + C_M + C_{FB} + C_{FT} + C_C + C_A + C_{TX} + C_{IP}$$

should be suitable changed as:

$$C_T = C_L + C_M + C_{FB} + C_{FT} + C_C + C_A + C_{TX} + C_{IP-NonTelecom} + C_{IP-TSP}$$

Magnitude of National consolidation of $C_{IP-NonTelecom}$ will decide what importance must be accorded by the country and how the progress on this access power utilization will be tracked.

8 Decomposed Problem Sets the Specific Focus and Clear Responsibility for Carbon Emission Reduction

This approach of decomposing the Carbon Emission problem with such holistic details creates opportunity to identify various aspects of the problems and sets the separate specific focus to address them. These specific aspects are

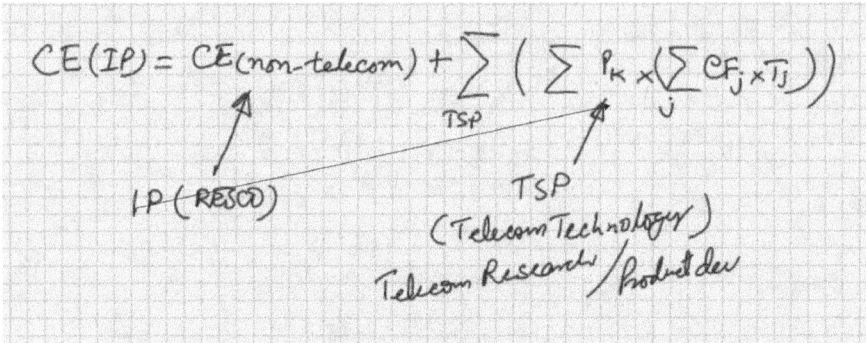

Figure 6 Estimation details sets responsibility to fix it.

related to the all the parts of the networks that forces to evolve and adopt all possible methods and procedure to reduce power consumption. It creates defined need for innovation in energy efficiency for future and of current telecom equipments. This will also fixed targets to utilize non-telecom power for alternate purposes and accounting the associated carbon emission as out of TSP budget. This will make RESCO model practical and in demand. The concerned policy making and business related actions would achieve speed. The Carbon emission breakup and the separate responsibility to address the problem is summarized in Figure 6.

By doing so, it also *sets the clear responsibility of TSP to motivate development and adoption of innovations carried out in energy efficiency and associated methods to reduce the power consumption and hence the saving on OPEX.*

A meaningful comprehensive frame work for the Carbon Emission Estimation and Reduction will prove helpful to evolve necessary policy changes to streamline the sector and help achieving the set target of carbon emission reduction.

With such approach and details of operation and understanding, it would be beneficial to set the Carbon emission reduction targets for each segment of the network that constitute the total carbon emission of a TSP as per the TRAI classification (section 2) and the recommended formula mentioned in previous sections as:

$$C_T = C_L + C_M + C_{FB} + C_{FT} + C_C + C_A + C_{TX} + C_{IP\text{-}NonTelecom} + C_{IP\text{-}TSP}$$

9 IOT Based System for Automatic Collection of Carbon Emission Parameters

The systems for automatic gathering the parameters required for estimating carbon emission would also be required to achieve the full objective of such a comprehensive solution for carbon emission estimation and reduction for establishing global leadership in this field. One of significant standardization delivery from GISFI's Technical Working Group on "Internet of Things (IoT)" as Application domain independent Generic Framework of IoT Based solutions is very appropriate candidate to be adopted for developing such system for Telecom Carbon Emission in India.

10 Conclusion

India has made a very good beginning towards achieving the Green Telecom objectives. Current regulatory definitions need refinements in terms of estimating the accurate Carbon emission from Telecom Network Equipments and setting up the reasonable and realistic targets of reduction in carbon emission.

The green telecom goal setting and the obligatory compliance will require necessary policy reforms that must be based on the sound study outcome carried out by Indian standardization organization with active participation from Telecom Service Providers, Telecom equipment providers, Technology Research & Development Organization and Government.

Some of the key Standardization and Policy formulation objectives are:

Evolution and adoption of the accurate carbon emission measurement estimation suitable for Indian telecom network.

Setup a national framework of information that contains the entire information related to carbon emission estimation from all the telecom service providers and Infrastructure providers. The common framework must be utilized for consolidating nationwide carbon emission estimates and study the necessary trend. This framework must relieve TSP to provide large reports as prescribed today. Regulator should also be able to draw necessary interpretation with regard to the verification of carbon emission data, estimation reports to check regulatory compliances and setting up the realistic and reasonable future goals.

Establishing a system for automatic information collection from the source of the information by building necessary technical mechanisms utilizing the recent advances in technology and their technical standards. The source of the information in this case is Infrastructure provider (IP) for providing

Switch-On/Off timing for various types of power sources and the consumed power. The source also includes inbuilt capability of each telecom equipment to assess and report the dynamic power consumption based on experienced traffic.

Adequate and immediate focus on achieving energy efficiency in the telecom network by following necessary Standards for measurements and metrics of energy efficiency, development of energy efficient technology building blocks, telecom products based on such building blocks and their adoption in the network under a national framework.

Access power generation at the Basestation sites is one major cause for escalated estimation of Carbon Emission in Indian Telecom Service provisioning that should be treated as separate problem to address. There should be a nationwide consolidation of the unutilized access power generated by IP for meeting the TSP requirement and a suitable policy should be formed for its utilization.

In order to provide adequate focus on the overall carbon emission reduction exercise, it would be necessary to set separate target of the carbon emission reduction for all the network segments separately.

References

[1] Telecom Regulatory Authority of India–Recommendations on Approach towards Green Telecommunications, 12 Apr, 2011
[2] GISFI TR GICT.105 V1.1.0 (2012-12): Metrics and Measurement Methods for Energy Efficiency
[3] GISFI TS GICT.101 V1.0.0(2013-02) Metrics and Measurement Methods for Energy Efficiency: Classification of Telecommunication Equipments; (Release 1)
[4] ETSI TR 102 530 V1.1.1 (2008-06) Environmental Engineering (EE) The reduction of energy consumption in telecommunications equipment and related infrastructure
[5] ETSI TR 102 532 V1.1.1 (2008-06) Environmental Engineering (EE) The use of alternative energy solutions in telecommunications installations
[6] ATIS-0600015.2009: Energy Efficiency For Telecommunication Equipment: Methodology For Measurement and Reporting - General Requirements (Baseline Document)

Biography

Krishna Sirohi has 25 years of experience from technology development leadership within both the telecom field and the defense sector. He is currently heading Impact Innovation in Technology & Business (i2TB) a technology incubation enabler unit of Samridhi Projects Private Limited (SPPL). i2TB is currently associated with few Technology Startup initiatives in their early or growth stages as well as with established organizations establishing their new technology businesses. He also heads Standards Committee of GISFI, the Global ICT Standardization Forum of India. GISFI is aligned with national policies and program in the field of Telecom, IT & Service based on emerging ICT infrastructure. He has been founding CTO of VNL, India's first initiative in private sector to develop sustainable rural telecommunication infrastructure using wireless technology. Prior to VNL he had served in Government's three major Indian R&D organizations in the field of Telecom and Defense Electronics. These organizations are CDOT, Bharat Electronics and Indian Telephone Industry.

Krishna has a Masters degree in Engineering (4 years Integrated course) from the Indian Institute of Science at Bangalore in year 1988 and BSC(H) in Physics from Delhi University.

He has been associated with technology development program of national importance. These programs includes India's first Large Capacity digital Switch design at ITI, India's first indigenous Command-Control-Communication-Intelligence (C3I) System for Indian Navy's Warfare Ships at BEL, India' first Mobile Network technology development program at CDOT and most innovative rural wireless solution at VNL.

He has represented in various international organizations and forums and contributed several internal forums and policy making exercises for domestic technology development programs.

He has been Vice-Chair of Special Interest Group (SSG) on "Beyond 3G Systems" of ITU-T and has leaded the standardization of Fixed and Mobile network convergence.

Email: president@i2tb.in, krishna.sirohi@gmail.com (alternate) Phone: +919899488800

Towards a Light Weight Internet of Things Platform Architecture

A. Sivabalan[1], M. A. Rajan[2] and P. Balamuralidhar[2]

[1]NEC India Pvt Ltd,Chennai, India; e-mail: sivabalan.arumugam@necindia.in
[2]Tata Consultancy Service, Bangalore, India; e-mail: rajan.ma@tcs.com,
balamurali.p@tcs.com

Received July 2013; Accepted August 2013

Abstract

This paper provides an overview of the activities of Internet of Things (IoT) work group in Global ICT Standardisation Forum for India (GISFI). Objective of this IoT WG is to identify potential standardization areas that can help proliferating the IoT technology and its applications that are relevant to India for the benefit of the society and businesses. The strategy chosen within this WG is to develop application independent generic IoT Framework with well-defined Reference Architecture to achieve interoperability between the various devices/application developed in multi-vendor scenario to achieve cost advantage and pass this advantage to the user group for its mass scale deployment and applicability. The requirements of the same are gathered through the study of various use cases.

Keywords: Internet of Things, Machine to Machine Communications, Service Platform.

1 Introduction

The Internet and World-Wide Web (www) has been a major driver of globalization and has promoted the convergence of electronic communications

Journal of ICT Standardization, Vol. 1, 241–252.
doi:10.13052/jicts2245-800X.12a8

and media services. Internet is continuing to become more pervasive, with the advent of low cost wireless broadband connectivity, by connecting to new embedded devices and handhelds. Further this evolution will continue to emerge as an "Internet of Things (IoT)" where the web will provide a medium for physical world objects to take part in interaction. This way the digital information technology can integrate the physical world to the online world to provide a common interaction platform.

IoT can be viewed as a global infrastructure for the information society, by interconnecting (physical and virtual) things based on, interoperable information and communication technologies towards providing advanced information services. Efficient exploitation of identification, data capture, processing and communication capabilities are integral requirements of IoT.

IoT is an integrated part of the future Internet that could be defined as a dynamic global network infrastructure with self-configuring capabilities linking physical and virtual objects through the exploitation of data capture and standard and inter-operable communication protocols.

This infrastructure includes existing and evolving Internet and will offer specific object-identification and addressing, sensor and connection capability as the basis for the development of independent and/or federated services and applications.

IoT promises to bring smart devices everywhere, from the fridge in your home, to sensors in your car; even in your body. Those applications offer significant benefits: helping users save energy, enhance comfort, get better healthcare and increased independence: in short meaning happier, healthier lives. But they also collect huge amounts of data, raising privacy and identity issues. For IoT to take off people need to feel a degree of comfort and control, and business need stability and predictability to invest. Therefore the issue of ethics and understanding needs, concerns and desires of people and businesses is so important.

These will be characterized by a high degree of autonomous data capture, context and event detection and transfer, network connectivity and interoperability at the protocol and semantic level with the provision of handling security and privacy concerns of the users and the data being communicated.

This paper presents the context and summary of activities in the IoT WG of GISFI highlighting some of the key contributions towards conceptualizing a light-weight service platform architecture. Section 2, discuss the background and scope of IoT WG activities. Section 3, summarizes major achievements of the IoT WG so far. Detailed discussion on GISFI IoT Baseline Light weight Architecture is presented in section 4. This is followed by a summary of a proposal for IoT Security in section 5. Last section briefly outline the planned scheme of future activities of the IoT WG.

2 Background and Scope

2.1 Background Information

As one of the fastest growth market for cellular technologies, India hasdemonstrated its appetite for technologies to revolutionize the life across all its diversity. The value, impact on daily life, and competition has helped the cell phone to reach the masses across all strata of the society.Urbanization of India is happening at a rapid pace. Migration from rural places to cities is ever increasing. One way to address this is to enhance the opportunities at villages to reduce the overburdening of the cities. Experience shows that this is not that easy. We need scalable architectures and funding models for building the city infrastructures to handle large populations.

Need and Challenges of IoT in India

There is a requirement for efficient systems for transportation, utilities, healthcare, safety & security, education, environment, governance and entertainment.Deployment of advanced ICT technologies would be affordable and cost effective considering the improved quality of life of citizens and enhanced GDP growth from the resultant productivity improvements.

A large percentage of Indian population is in rural are as and there is a constant drive for addressing those sections of the society. Majority of themanpower is spent on agriculture and farming. Many cases they are resource-constrained in terms of water, energy, fertilizers, and market opportunities.

Systems for monitoring and improving the efficiency of resource utilizationwill be highly beneficial. Health care is another area of attention here where remote monitoring can enable the skilled doctors in cities to extend their services to villages.

There are many challenges for successful adoption of IoT in India.

Scalability: As the size of the systems tends to be large in size, the solutions should be scalable. Also many times the deployments happen in stages and the architecture should be able to scale-up incrementallywithout taking too much overhead.

Affordability of products and services: Affordability is one of the major aspect for success. It may not be low cost always, but the right cost for a specified target group with a clear business case or cost benefit. Standardized platforms, tools and manufacturing processes can bring thecost down with increased volumes.

Integration with Legacy systems: Since there are no widely deployed IoT applications, there may not be any major challenge with legacytechnologies in that space. However there may be legacy devices andsystems which are not amenable for new standardization and need to coexist.

Robustness: While there is a pressure for low cost, there is a strong demand for robustness and reliability of products and services. One of the approaches for addressing this is to build upgradable/disposable systemswhich take care of current requirements and strip of the low priority features to reduce cost. A Robust solution will get a buy-in even if it isless sophisticated.

Social and Cultural Sensitivity: Social response to an IoT application has many aspects. It can have cultural, linguistic, geographic, political dependencies for the acceptance. Help of awareness and regulations areto be explored for the success of large scale social applications

Over the last three years, the IoT WG along with stakeholders has put forth significant efforts to carry out detailed study of the IoT/ Machine to Machine (M2M) standards landscape,regulations and Indian needs on IoT. Major participating industrial organizations in this work group include Tata Consultancy Services (TCS),NEC, Ericsson, Cisco, I2TB-SPPL, Wirefreecom, ILS Technologies andOPC Foundation. TCS holds the current chair of the workgroup and Ericsson is the Vice Chair. There have been more than thirteen WG meetings till date [13].

2.2 Scope and Objectives

Objective of this IoT WG is to identify potential standardization areas that can help prolife rating the IoT technology and its applications that are relevant to India for the benefit of the society and businesses.

The strategy chosen within IoT Working Group of GISFI is to develop application independent Generic IoT Framework with well-defined Reference Architecture to achieve interoperability between the various devices/ application developed in multi-vendor scenario to achieve cost advantageand pass this advantage to the user group for its mass scale deployment and applicability.

The examples of such IoT applications include: fleet management, smart metering, home automation, e-health, e-agriculture, smart cities, smart manufacturing,environment and natural resources management etc.

3 IoT WG Achievements

Achievements of IoT WG so far are in the following aspects:

- Study of the requirements of IoT for use cases that are relevant to India.These studies resulted into a number of technical reports.
- Reference Architecture for IoT for organizing and synergizing various-standardization requirements of IoT.
- A light weight security scheme for IoT applications

3.1 Use case scenarios presented

There are several use case scenarios presented in workgroup that motivates the requirement of a common standard across multiple application domains,with relevance to India. These use cases were submitted by different participants and some of the major ones are:

  (i) Agriculture monitoring [1]
 (ii) mHealth [7,11], Connected Health Care [3,4]
(iii) Landslide detection [6]
(iv) Food Supply Chain Management [8]
 (v) Security,Surveillance [9]
(vi) Smart metering and control [10]
(vii) Smart Cities [12]

4 GISFI IoT Baseline Light weight Architecture

In this section, we present more details on a reference architecture for IoT which discusses a functional architecture of the IoT stack [2]. The scope of the architecture is from sensors/devices to applications. The IoT stack is expected to capture the heterogeneity of devices and communication protocolsat the lower layer and to provide uniform interfaces to the upper layers.

Objectives of formulating reference architecture are multi-fold and are explained as follows:

- It identifies major reference points / interface points which can be considered for standardization to encourage interoperability of products and-services from multiple stake holders.
- It helps in explaining various IoT use case scenarios and gathering respective requirements of these interfaces

- Developing consistency in information exchange and contributions from multiple participants of this standards development effort

4.1 Requirements

The reference architecture should be able to identify key architectural components and interfaces so that multiple stake holders providing products and there are several legacy devices, technologies and standards existing at thelower layer of TCP/IP layering in the immediate vicinity of objects and devices. The architecture should support them optimally.

In addition to the above requirements, some of the India specific requirements that need to be considered are discussed below.

India has a huge population with diversity in terms of geography, culture and other socio-economic factors. The architecture should support the IoT services that can scale, at the same time support multiple levels of sophistication in terms of technologies and devices coexisting.Even within a single application domain such diversity needs are to be supported.

The nationwide coverage of candidate networks for IoT Core connectivityis not uniform, rather patchy. This unreliability in communication needs to be taken into account and suitable measures for robustness of IoT connectivity needs to be incorporated. Moreover the large numberof IoT devices can lead to unexpected peak requirements and that may bring down the network due to congestion and limited capacity.This requires advanced management techniques for the IoT network and services.

Affordability of IoT services is an important aspect where low cost technologies should form the part of the baseline services. In certain large scaleapplications, massive deployment of sensing devices may not be feasible dueto cost considerations.

In consideration with such scenarios a human enabled sensing mechanism is included in this scope where a device such as mobile phone carried by a person is used for communication with objects/persons and integrate with the IoT services. Further the architecture should enable thereduction of cost of intellectual properties and encourage IPR development and competition in the country towards better control of affordable services.

4.2 IoT Architecture

The scope of the reference architecture discussed here (refer Fig. 1) includes the entire cycle of IoT applications, from sensing to application services [2].

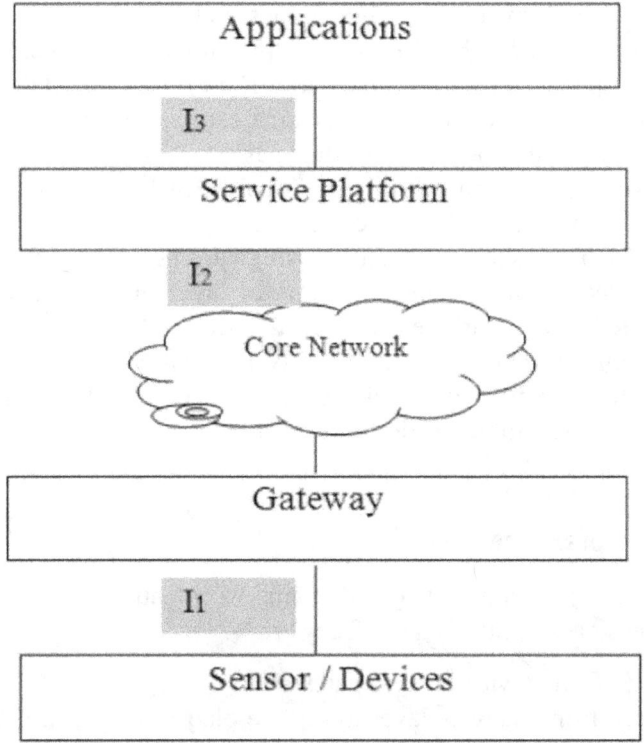

Figure 1 IoT reference architecture.

It is partitioned into three layers namely Device layer, Gateway layer, and Service Platform layer. This paper also discusses issues involving IoT core network asfollows.

IoT Device Layer: IoT devices are included in this layer. This layer consists of individual sensors, network enabled objects and capillary networks consisting of datasources that are near to the physical environment. It includes heterogeneous devices (including sensors and actuators) supporting diverse communication standards such as Zigbee, ZWave, ANTS and Wi-Fi, etc.

IoT Gateway Layer: This layer consists of IoT gateways. The substantial heterogeneity of devices and technologies hosted by the device layer is abstracted using gateways that can provide a more uniform interface to IoT service platform layer. It is also possible that a capable device can implement both IoT device andgateway layer/functionality into a single physical entity and connects to the IoT service platform layer through the core network

IoT Service Platform Layer: This defines and provides different IoT service abstractions that can be used by multiple applications. There can be a set of platform services from the IoT platform infrastructure. Further the same framework can be extended to application services where some of the reusable application components are available as services.

IoT Core Network: The physical entities involved in the above three layers need suitable communication infrastructure for information exchange. While the device layer addresses this requirement using various legacy technologies which are outof scope for this paper, the gateway layer and service platform layer are expected to be connected over an IoT Core / Backbone network. The IoT Coreis envisaged to be predominantly an IP based network and that is in line with the vision of IoT. This IP connectivity could be supported over multitudes oftelecommunication infrastructures such as DSL, Cellular networks (2G, 3G, 4G) etc.

4.3 Interface Reference Points

Further to the identification of major domains we identify three reference points at the interfaces of these layers. They are

> I1: Interface from device layer to gateway layer,
>
> I2: Interface from gateway layer to service platform layer through IoT core network
>
> I3: Interface from service platform to layer specific vertical applications

Each of these interface points will benefit from a standardized information exchange because of the diversity of devices, manufacturers, serviceproviders, and service consumers involved. Each of these interface points are expected to support a set of specialized capabilities which may form aset of standardized adapters designed for the purpose. These adapters may implement existing protocols or needs new developments or extension based on the requirements gathered from various IoT use cases.

5 IoT Security

For secure communication and authentication between the devices, cryptography is an essential tool. In general, cryptography with PKI is very popular, wherein the certificates containing the public and private keys are issued to the users and users using these keys encrypt and decrypt their data and thus security is enabled. The same scheme is not viable for IoT scenario. The crypto system for IoT security has the following requirements. The requirements are broadly classified into two categories to ensure Confidentiality, Integrity and Availability.

- Communication Security.
- Storage Security.

5.1 Communication Security

The communication security deals to handle the security for the IoT communications. It has following requirements:

- Secure Device to Device Communication: Here the cryptographic technique should enable secure IoT communication across I1, I2 and I3 interfaces. The cryptographic technique should be lightweight.
- No Public Key Infrastructure (PKI):Number of devices in IoT is very large, maintaining keys at PKI is not feasible. So cryptographic technique with no PKI is very essential
- Certificate less cryptography: IoT demands less infrastructure for security, hence the requirement here is to design crypto systems with no certificates.
- No Key Exchange: One of goal of IoT is to reduce number of control/communication messages. This is relevant to secure communications also. Since device to device communications in IoT are very prevalent, to minimize the message overheads, crypto system should support no key exchanges.
- Anonymity: In IoT, prominently in the area of disaster management/mission critical operations, to avoid attacks from the intruders, anonymity is required. So it is desirable that crypto system should support security as well as anonymity also.
- Variable security requirement: Handling of heterogeneous devices with differing protocols.
- Group Communication: In IoT, very often the Application (layer 4) requires data from different devices and also need to control them

5.2 Storage Security

IoT is poised to be the largest source of data generator from a wide range of diversified applications. How to store and archive the data at devices, platform, and service providers robustly is a big challenge. Thusdata should be delivered or distributed to the intended users/applications only in a secured and anonymized way to satisfy privacy requirements. Therefore it is desirable to have a lightweight crypto system for secure collection, storage, and archive and distribution of data.

Hence a scalable, lightweight, with no infrastructure, certificate less and no key exchange cryptographic technique is essential for secure IoT communication. Thus using IBE, we can envisage the requirements of the secure communication for IoT.

5.3 Applicability of Identity based Encryption for IoT

Identity based Encryption (IBE) is a secure certificate-less cryptography scheme, wherein the devices can generate the public key of the other devices by using publicly known identity of the devices such as device's, owner's id, mac-id, etc. and encrypt the message with this public key and on the other hand the device which receives the encrypted message shall decrypt the message using its' private key (obtained during device registration/bootstrapping in IoT).Key revocation is an issue in IBE that needs to be addressed. The techniques such as Lattice based IBE and Attribute Based IBE schemes to enable key revocation.

There are a few IBE based standards which are still evolving. IEEE has a draft standard on IBE cryptography IEEE P1363, which is very generic and not specific to IoT/M2M communication.

For M2M and IoT, the existing schemes proposed by EuropeanTelecommunications Standards Institute (ETSI) based on IBE are through key exchange algorithms (Diffie–Hellman algorithm) and authentication using IBAKE algorithm and uses Weil paring in cryptography. The proposal made in the Working Group (WG) is to use IBE cryptographic schemes with or without key exchanges along with anonymity, authentication and signature schemes.

6 Going Forward

Elaboration of the interfaces specified in the GISFI IoT reference architecture that addresses the major requirements of various use-case scenarios is the work in progress in the work group. The specification is oriented towards the specification of a Light Weight IoT/M2M Framework with a focus on the impact use cases relevant to India. It requires the identification of baseline requirements those are categorized into mandatory, desirable, and optional.The inputs for requirements are sourced from various GISFI documents (use case scenarios, framework document), and international standards such as ETSI. This selection need to be guided by the India specific challenges.

References

[1] GISFI_IoT_20110680, "Agriculture Application Requirements", June 2011
[2] GISFI_IoT_201206218, "Internet of Things Reference Architecture", June 2012.
[3] GISFI_IoT_201206221, "Indian relevance to Healthcare Service Delivery mechanism based on Generic IoT Framework", June 2012
[4] GISFI_IoT_201206222, "Healthcare Service Delivery mechanism based on Generic IOT Framework", June 2012.
[5] GISFI_IoT_201206227, "IoT Service Capabilities", June 2012
[6] GISFI_IoT_201206228, "Landslide Detection Use-case", June 2012
[7] GISFI_IoT_201203179, "mHealth Use Cases", March 2012.
[8] GISFI_IoT_201203180, "Food Supply Chain Management (FSCM) Use Cases", March 2012.
[9] GISFI_IoT_201203181, "Surveillance Security System Use cases", March 2012.
[10] GISFI_IoT_20110677, "Privacy Requirements of User Data in Smart Grids", June 2011
[11] GISFI_IoT_20110687, "Frame-work Document for TR on e-Health Use Case", June 2011
[12] GISFI_IoT_201209291, "Smart City Usecase", Sep 2012.
[13] GISFI_IoT_201212335, "Security Requirements and Proposal for IoT", Dec 2012.[13] GISFI Meeting Documents: IoT URL: http://gisfi.org/workinggroups.php?wg=IOT [Last accessed: 12th August, 2013]

Biographies

Sivabalan A is working as a Manager – Research in NEC Mobile network Excellence Centre(NMEC), at NEC India Pvt Ltd, Chennai India. He has 13 years of experience in Industrial and academic research. Prior to NEC, he was the Associate Scientist in Industrial Communication Research Group, INCRC at ABB Global Services and Industries Limited, Bangalore, India. He received Ph.D in Electrical Engineering from Indian Institute of Technology, Kanpur (IITK) and also carried-out Postdoctoral research with Physical and Digital Realisation Research Group at Applied Technology Research Centre, Motorola India Research Lab, Bangalore India. He has around 30 Journals and International Conferences publications.His current role includes representing NEC in Global ICT Standards forum of India (GISFI). He research interest includes Free space Optical Communication, 802.xx Physical Layer, Device communication and Integration.

Rajan M A, who obtained B.E., M.Tech.and PhD Degrees in Computer Science and Engineering and Mathematics, M.Sc., M.Phil in Mathematics. He, is currently employed as a Scientist at TCS Innovation Labs, Bengaluru. He also worked as a Scientist in Indian Space Research organization from December 2000 to September 2005 and was actively involved in realization of several spacecrafts.. Apart from industrial experience he is also working as a visiting academic faculty in SJCIT, Chikkaballapur and also in UVCE, Bengaluru for over 12 years. In an overall he is having 13+ years of both industry and academic experience in the field of computer science and engineering. His research activities involves Cryptography, information security, Privacy preserving techniques, Computer Networks, Cross layer design, Number Theory, Graph Theory, Combinatorics, coding theory and Functional Analysis. He has National and International patents and also published several research papers in national and international conferences and journals.

Balamuralidhar P is a Principal Scientist and Head of TCS Innovation Lab at Tata Consultancy Services Ltd (TCS), Bangalore. He has obtained Bachelor of Technology from Kerala University and Master of Technology (MTech) from IIT Kanpur. His PhD is from Aalborg University, Denmark in the area of Cognitive Wireless Networks. Major areas of current research include different aspects of Cyber Physical Systems, Sensor Informatics and Networked Embedded Systems. Before TCS his research careers were with Society for Applied Microwave Electronics Engineering & Research (SAMEER) Mumbai and Sasken Communications Ltd Bangalore.

He has over 25 years of research and development experience in Signal Processing, Embedded Systems and Wireless Communications. He has over 60 publications in various international journals and conferences and over 20 patent applications. Balamuralidhar was the leading TCS participation in two EU FP6 research consortium projects namely My Adaptive Global NET (MAGNET) and End to End Reconfigurability (E2R) in the area of next generation wireless communications. He is also contributing to TCS participation in National bodies like Broadband Wireless Consortium India (BWCI), Global ICT Standards for India (GISFI). In GISFI he is chairing the Internet of Things Workgroup.

Interoperation among IoT Standards

Soma Bandyopadhyay, P. Balamuralidhar and Arpan Pal

Innovation Lab, Kolkata, India
Tata Consultancy Services Ltd, India
email: {soma.bandyopadhyay, balamurali.p, arpan.pal}@tcs.com

Received July 2013; Accepted August 2013

Abstract

This article presents generic IoT (Internet of Things) reference architecture proposed by standard organization. It compares different IoT reference architectures proposed by different standards institutes, specially emphasizing lightweight requirements of IoT reference architecture. These architectures define multiple interfaces. Functionalities of proposed interfaces are compared here to enable an understanding regarding the scope of interoperation among different IoT standards. Finally, future scopes for standardization in IoT are presented.

Keywords: IoT, Standardization body, lightweight.

1 Introduction

Internet of Things (IoT) is fundamentally network of networks with Internet as backbone. It associates diverse sensors, actuators, computing system to provide intelligent services to human society. It comprises sensing network with different constrained sensors. These sensors detect environmental condition, send this sensing information to backend computing system over Internet. Backend interprets sensed data, sometimes it is done by sensor/sensor-gateway

Journal of ICT Standardization, Vol. 1, 253–270.
doi:10.13052/jicts2245-800X.12a9
© 2013 *River Publishers. All rights reserved.*

locally, takes decisions, and routs these decisions to local applications, actuators or to other user devices like mobile phones which are also constrained in nature.

[1] Discusses different visions of IoT. In [2] functionalities and components of middleware for IoT architecture are presented.

Figure 1 depicts the said communication components of IoT. It is evident from above figure that IoT architecture needs to address the limitations of constrained network acting as networks of data producers as well as data consumers.

The very purpose of IoT is to make things smart [14]. Different use cases and technologies of IoT are depicted in Figure 2 below.

Diverse technologies, software, applications are building blocks of IoT. Many standard organizations are putting effort to define IoT reference architecture emphasizing machine to machine (M2M) communication which is pragmatic to define functional modules of IoT architecture.

ETSI (European Telecommunications Standards Institute) [3], ITU-T (International Telecommunication Union) [8], TIA (Telecommunications Industry Association) [9], OMA (Open Mobile Alliance) [12], GISFI (Global ICT Standardization Forum for India) [11], CCSA (China communication standard association) [10], CASAGRAS (coordination and support action for global RFID-related activities and standardization) are some examples of such standard organizations working on IoT and M2M communication. Figure 3 depicts different standard bodies working in different areas and use-cases of IoT.

A state-of-the-art in M2M communications, in terms of standardization bodies, research projects, protocols, etc. along with application programming interfaces (API) for endpoints of IoT communication network is analyzed in [13]. Figure 4 shows association among ETSI, OMA, TIA etc. Diversity is key factor of IoT therefore interoperation is a major challenge to achieve this diversity. Different categories of interoperability like semantic, syntactic, technical and organizational are presented in [4]. Figure 5 represents different

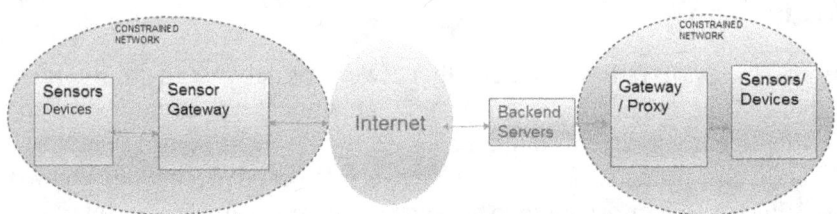

Figure 1 IoT as network of networks with Internet as backbone.

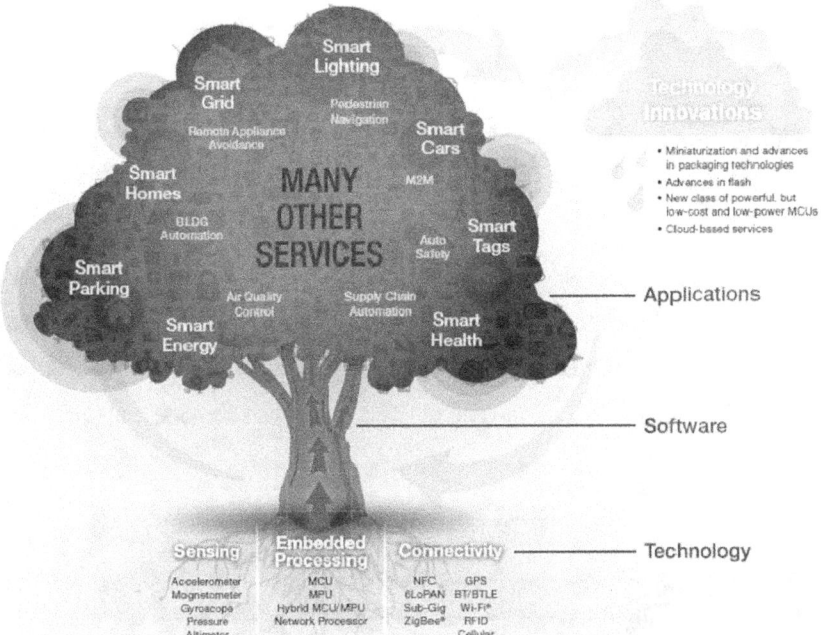

Figure 2 Different services, technologies of IoT [14].

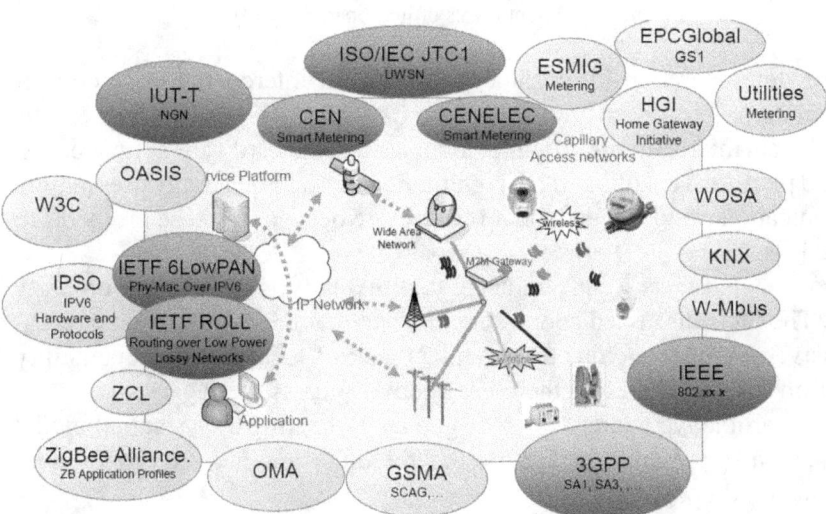

Figure 3 Different SDOs and industrial organizations working on different segments of IoT communication network and IoT applications [11].

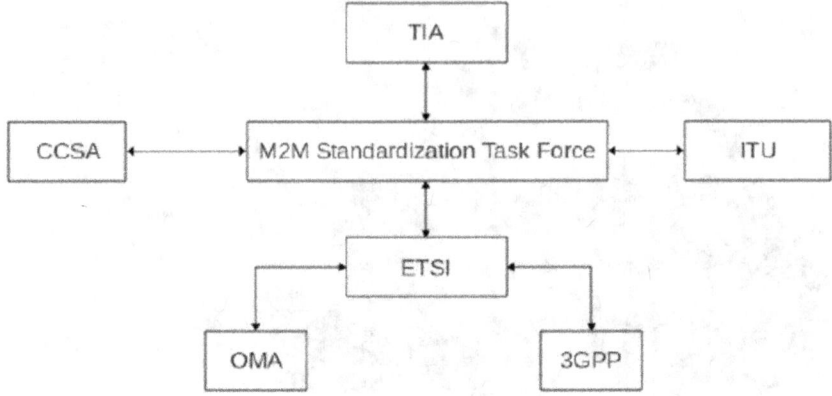

Figure 4   Collaboration among different standard bodies working on M2M [13].

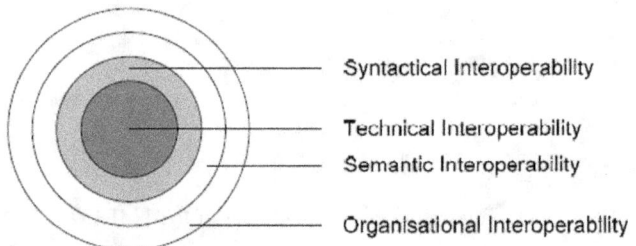

Figure 5   Different levels of interoperability [4].

levels of interoperation by considering technical interoperation as core of interoperation.

Technical interoperability defines association with hardware/software, systems and platforms enabling M2M communication. This generally uses mainly communication protocols and the infrastructure needed for those protocols to operate [4].

Above stated facts clearly indicate that diversity is an inherent property of IoT. The facts discussed above also indicate that there are 1) lack of comprehensive reference architectures, and 2) lack of technical interoperability evaluation scopes. Therefore there is a need to address these gaps.

In this article we address the first gap by presenting a conceptual model of IoT, and also presenting a generic IoT reference architecture proposed by GISFI standard body.

IoT reference architecture, its functional modules and interfaces defined by different standard bodies mainly ETSI and GISFI are compared here. Further based on these comparisons we discuss scope of technical interoperation aspects among said standard-bodies to address the second gap. None of above literatures perform an extensive study in this regard.

2 IoT Reference Architecture

In this section we present conceptual model of IoT, and its reference architecture proposed by different standard body. IoT architecture constitutes mainly four layers like sensor or thing layer, network layer, service, and application layer.

Both ETSI, and GISFI reference architecture for M2M and IoT support this conceptual model. Figure 6 depicts the conceptual model of IoT.

2.1 ETSI Reference Architecture

ETSI high level reference architecture possesses two domains as stated below [15].

1. **The device and gateway Domain.**

It is composed of the following elements:

M2M device: A device that runs M2M application(s) using M2M service capabilities. M2M devices connect to network domain using a direct

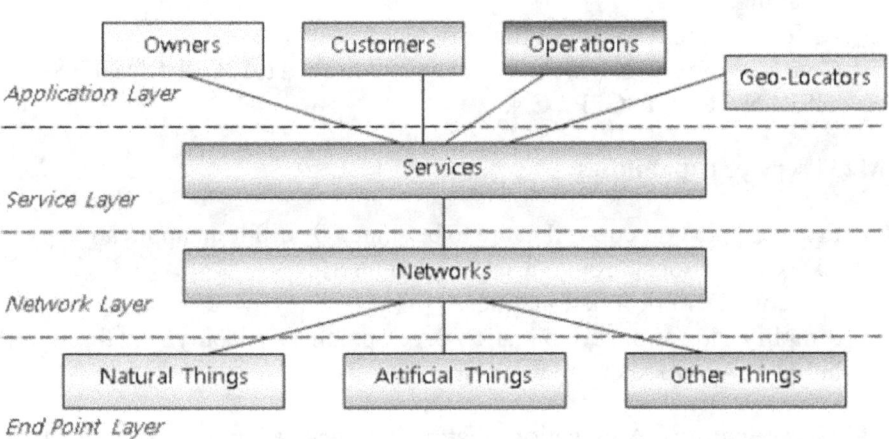

Figure 6 Conceptual model of IoT [5].

connectivity, or using gateway as a network proxy. M2M devices may be connected to the networks domain via multiple M2M gateways.

M2M area network: provides connectivity between M2M Devices and M2M Gateways. Examples of M2M Area Networks include: Personal Area Network technologies such as IEEE 802.15.1, Zigbee, Bluetooth, IETF ROLL, etc. or local networks such as PLC, M-BUS, Wireless M-BUS and KNX.

M2M gateway: A gateway runs M2M Application(s) using M2M service capabilities. The gateway acts as a proxy between M2M devices and the network domain.

2. **The network domain**

It is composed of following elements:

Access network: Network which allows the M2M device and gateway domain to communicate with the core network. Access Networks include (but are not limited to): xDSL, HFC, satellite, GERAN, UTRAN, eUTRAN, W-LAN and WiMAX.

Core network:

It provides following:

> IP connectivity at a minimum and potentially other connectivity means.
> Service and network control functions.
> Interconnection (with other networks).
> Roaming.
> Different Core Networks offer different features sets.
> Core Networks (CNs) include (but are not limited to) 3GPP CNs, ETSI TISPAN CN and 3GPP2 CN.

M2M service capabilities:

> Provide M2M functions that are to be shared by different applications.
> Expose functions through a set of open interfaces.
> Use core network functionalities.
> Simplify and optimize application development and deployment through hiding of network specificities.

M2M applications: Applications that run the service logic and use M2M Service capabilities accessible via an open interface.

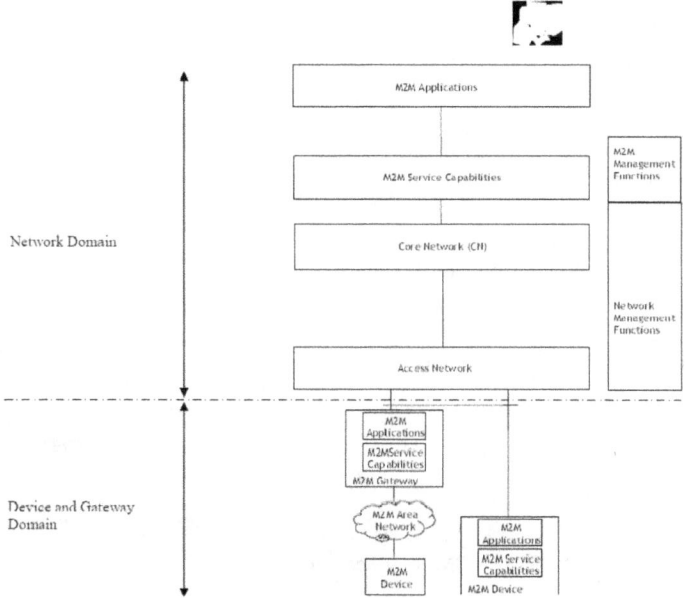

Figure 7 ETSI Reference Architecture [15].

ETSI M2M architecture has following interface reference points [15]. Reference points are described below based on [15].

mIa Reference Point:
Allows a network application to access M2M service capabilities in the network domain. The mIa reference point shall comply with the specification [16].

dIa Reference Point:
Allows a device application residing in an M2M device to access the different M2M service capabilities in the same M2M device or in an M2M gateway; allows a gateway residing in an M2M gateway to access different M2M service capabilities in the same M2M gateway. The dIa reference point shall comply with specification [16].

mId Reference Point:
Allows an M2M service capabilities residing in an M2M device or M2M gateway to communicate with M2M service capabilities in network Domain and vice versa. mId uses core network connectivity functions as an underlying layer. The mId reference point shall comply with the specification [16].

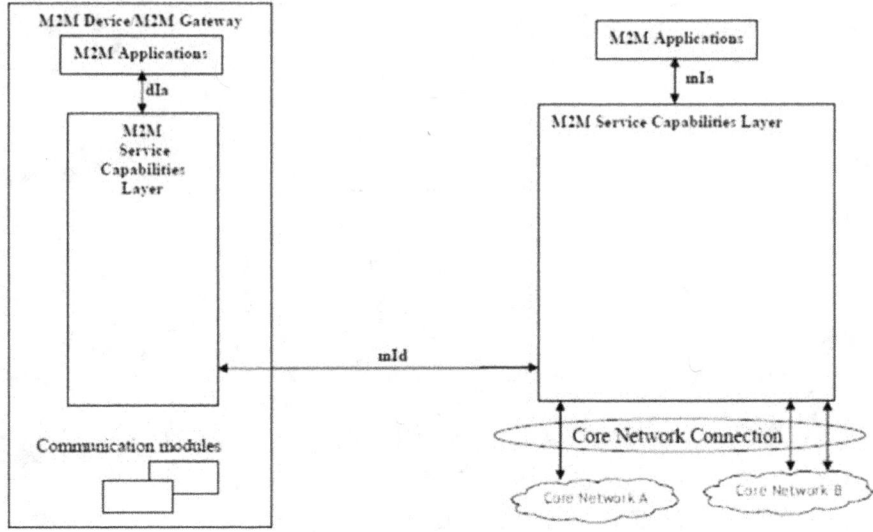

Figure 8 ETSI M2M interfaces (mId, mIa, dIa) [15].

Figure 8 represents interfaces mIa, mId and dIa defined by ETSI.

2.2 GISFI IoT Architecture

In this section we present generic IoT reference architecture proposed by GISFI –IoT- WG. The IoT reference architecture of GISFI follows conceptual model presented in figure [6].

GISFI IoT reference architecture is partitioned into four layers namely device layer, gateway layer, service platform layer and application layer. Different layers are described in brief below following the GISFI – IoT-contribution [6].

- **IoT Device Layer**

IoT devices are included in this layer. This layer consists of individual sensors, network enabled objects and capillary networks consisting of data sources that are near to the physical environment. It includes heterogeneous devices (including sensors and actuators) supporting diverse communication standards such as Zigbee, ZWave, ANTS andWi-Fi, etc.

- **IoT Gateway Layer**

This layer consists of IoT gateways. The substantial heterogeneity of devices and technologies hosted by the device layer is abstracted using gateways that can provide more uniform interface to IoT service platform layer. It is also possible that a capable device can implement both IoT device and gateway layer/functionality into a single physical entity and connects to IoT service platform layer through the core network.

- **IoT Service Platform Layer**

This defines and provides different IoT service abstractions that can be used by multiple applications. There can be a set of platform services from the IoT platform infrastructure. Further the same framework can be extended to application services where some of the reusable application components are available as services.

The physical entities involved in the above three layers need suitable communication infrastructure for information exchange. While device layer addresses this requirement using various legacy technologies which are out of scope for this document, the gateway layer and service platform layer are expected to be connected over an IoT Core / Backbone network. The IoT Core is envisaged to be predominantly an IP based network and that is in line with the vision of IoT. This IP connectivity could be supported over multitudes of telecommunication infrastructures such as DSL, Cellular networks (2G, 3G, 4G) etc.

- **IoT Application Layer**

This layer consists of different IoT applications.

GISFI IoT architecture has following interface reference points at the interfaces of these layers stated above.

- I1: Interface from device layer to gateway layer,
- I2: Interface from gateway layer to service platform layer through IoT core network
- I3: Interface from service platform to layer specific vertical applications

Each of these interface points will benefit from a standardized information exchange because of the diversity of devices, manufacturers, service providers, and service consumers involved. Each of these interface points are expected to support a set of specialized capabilities which may form a set of standardized adapters designed for purpose. These adapters may implement existing protocols or needs new developments or extension based on requirements gathered from various IoT use cases. Figure 9 shows the GISFI IoT interfaces. Functionalities of these interfaces are summarized in Table 1 [6].

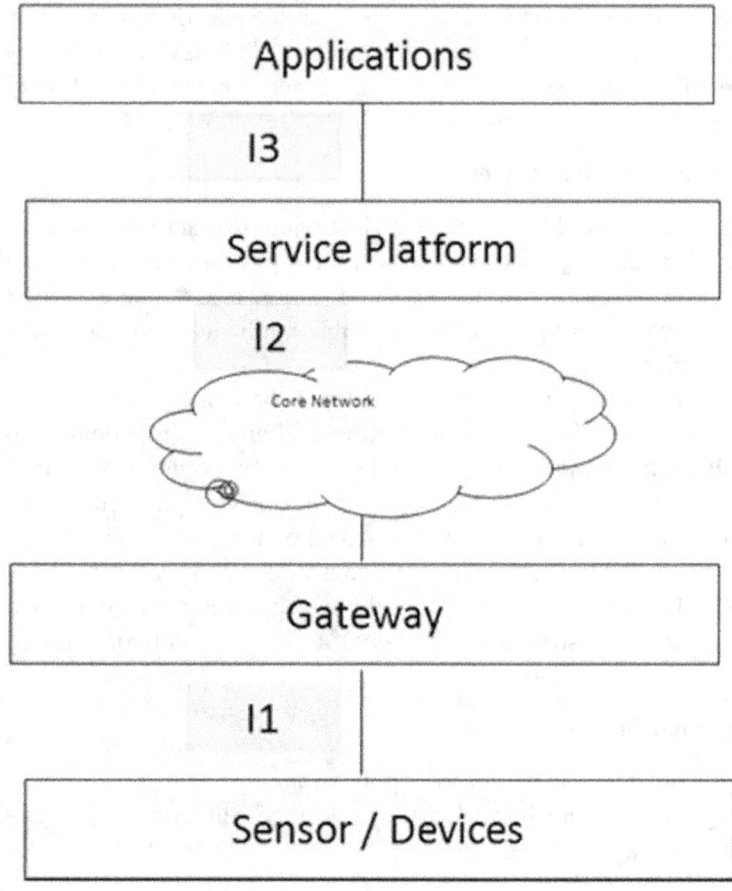

Figure 9 GISFI IoT interfaces (I1, I2, and I3) [6].

2.3 ETSI and GISFI standard-organization comparison and scope of their interoperation

This section presents scope of technical interoperation between ETSI and GISFI standards.

Comparison of interfaces proposed by GISFI– IoT–WG and ETSI-M2M-WG are presented in table 2 below based on the GISFI contribution [7]. M2M and IoT reference architecture of ETSI and GISFI indicating interfaces are presented in Figure 10.

Table 2 represents comparisons of interfaces of ETSI M2M and GISFI IoT reference architecture. GISFI IoT architecture proposes three (I1, I2,

Table 1 Functionalities of IoT interfaces proposed by GISFI IoT reference architecture

Interfaces	Capabilities
I1	This is the reference interface between the IoT device and IoT gateway. I1 will accommodate co-existence of multiple legacy link level and sensor network standards. A unified data interchange format between sensors and gateway can be a focus here.
I1a	I1a is the interface that has the capability of handling the data path between devices and gateway.
I1b	Device specific management functions such as sensor sampling configuration, security settings, device registration, device health check, firmware upgrade etc will be done through this interface.
I2	This is the reference interface between the IoTgateway and IoT-service platform. Service platform is an application middleware providing platform services to build domain specific applications. This has a data path as well as management path. The connectivity is provided using IoT Core Network which is predominantly any IP capable core network infrastructure such as xDSL, 3GPP, 3GPP2.
I2a	It exchanges the data that includes various sensor observations, aggregates from the IoTgateway to the IoT Service Platform.
I2b	This interface takes care of the gateway management functions including security/authentication configuration, firmware upgradation, application download, health monitoring etc.
I2c	This interface takes care of communication between gateways to enable mobility, resilience, and scalability.
I3	This interface provides a uniform access of various IoT services to the domain specific vertical applications. The applications may run at different physical entities but they need to have the IoT services access from multiple service providers.
I3a	This interface exposes the access to various data services
I3b	This interface provides access to various management and administrative services. This includes user management, device registration, storage management, IoT application store, access control, privacy etc.
I3c	It enables IoT service platform to communicate with another IoT service platform towards scalability and data exchange with peer IoT systems.

I3) interfaces. ETSI also defines three interfaces (dia, mid, mia). Both standards support data and management functionalities. However I2c and I3c defined by GISFI support self-looping among similar gateways and service platforms these functionalities are missing in reference architecture defined by ETSI.

Technical-interoperability of IoT framework from different standards is achievable as long as standards abide by the concept of three layered

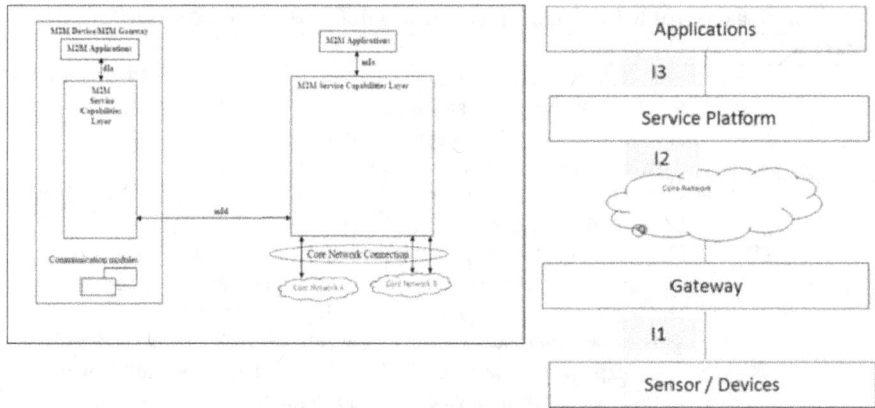

Figure 10 IoT reference architecture: ETSI (left) and GISFI (right) showing interfaces.

architecture (sensor, core/backbone network, application/services) and conceptual IoT architecture model depicted in figure 6. However semantic, syntactic interoperability are achievable by performing mapping among different groups (like mandatory, optional) of attributes of interfaces and different APIs (application programming interface), out of scope of this article.

3 Conclusion

Establishing interoperation across various IoT reference architectures defined by different standard organizations specifically to make them technically interoperable is explored in this document. The architectures consisting multiple layers like sensors, core-network, service platform and applications possess technical interoperability. We have elaborated architecture and functionalities of reference interfaces proposed by GISFI. Elaboration of interfaces specified in GISFI IoT reference architecture addressing major requirements of various use-case scenarios is work in progress in that work group. We have compared functionalities of proposed interfaces of reference architecture defined by ETSI and GISFI. Establishing low overhead secured path with unique naming and addressing mechanism is still an open area for standardization. We believe bringing different standards semantically and syntactically interoperable by addressing main requirements of IoT framework like diverse and distributed computing environment, low overhead communication medium, support of critical events, and human machine interaction is another research area and candidate for standardization.

Table 2 Comparison among different interfaces of GISFI and ETSI.

Interfaces	Type	Requirement	Capabilities	Standard body
Device and gateway (Sensor domain)			This is reference interface between the sensor device and gateway. It accommodates co-existence of multiple legacy link level and sensor network standards. A unified data interchange format between sensors and gateway can be a focus here.	ETSI(dia), GISFI (I1),
	Data	Required device specific protocol, low overhead and optimized way of energy and bandwidth usage, datapublishing capability. Configuration attributes are required for exchanging information with sensors.	I1a is interface that has the capability of handling the data path between devices and gateway.	ETSI, GISFI
	Management	Required to handle configuration parameter into device specific format- like XML/SensorML. Exchanges device registration, capability negotiation, device configurations, etc.	Device specific management functions such as sensor sampling configuration, security settings, device registration, device health check, firmware upgrade etc will be done through this interface.	ETSI, GISFI
Gateway and service platform interfacing core network			This is reference interface between gateway and service platform. It has a data path as well as a management path. The connectivity is provided using IoT/M2M core network which is predominantly anIP capable core network infrastructure such as xDSL, 3GPP, 3GPP2, etc.	ETSI(mid), GISFI (I2)
	Data	Required device specific protocol, low overhead and optimized way of energy and bandwidth usage capable of data publishing – example CoAP(constrained application protocol) [17]	It exchanges data that includes various sensor observations, aggregates from gateway to service platform.	ETSI, GISFI

Table 2 Continued

Interfaces	Type	Requirement	Capabilities	Standard body
	Management	Required a low overhead reliable, and secured communication channel, supporting large payload size data transfer (example: block mode of CoAP)	This interface takes care of gateway management functions including security/authentication configuration, firmware up-gradation, application download, health monitoring etc.	ETSI, GISFI
	Self-loop Similar Gateways	Required low cost communication channel, group communication (broadcast, multicast)	This interface takes care of communication between gateways to enable mobility, resilience, and scalability.	GISFI, not present in ETSI
Service platform to applications			This interface provides uniform access of various IoT services to domain specific vertical applications. The applications may run at different physical entities but they need to have IoT services access from multiple service providers.	ETSI (mia) GISFI (13)
	Data	Data, from various applications to the service platform needs to be exchanged, required generic API set.	This interface exposes the access to various data services	ETSI, GISFI
	Management		This interface provides access to various management and administrative services. This includes user management, device registration, storage management, IoT application store, access control, privacy etc.	ETSI, GISFI
	Intra-layer communications		It enables Service Platform to communicate with another service platform towards scalability and data exchange with peer IoT systems.	GISFI, not present in ETSI

References

[1] http://numenor.cicese.mx/cursos/CMU/atzori-iotsurvey.pdf

[2] Soma Bandyopadhyay, MunmunSengupta, SouvikMaiti, Subhajit Dutta "A Survey of Middleware for Internet of Things" CoNeCo 2011, Ankara, Turkey, June 26 - 28, 2011.

[3] http://www.etsi.org/deliver/etsi_ts/102600_102699/102690/01.01.01_60/ts_102690v010101p.pdf

[4] http://www.etsi.org/images/files/ETSIWhitePapers/IOP%20whitepaper%20Edition%203%20final.pdf

[5] GyuMyoung Lee, "Standardization on the Internet of Things and Applications for Smart Grid", 2nd u-World Congress, Dalian, China, August 2012

[6] http://www.gisfi.org/wg_documents/GISFI_IoT_201206218.doc

[7] http:// www.gisfi.org/wg_documents/ GISFI_IoT_201106103.doc

[8] http://www.itu.int/en/ITU-T/Pages/default.aspx

[9] http://www.tiaonline.org/standards/procedures/manuals/scope.cfm#TR50

[10] http://ccsa.org.cn/english/tc.php?tcid=tc10

[11] http://www.gisfi.org/index.php

[12] http://www.openmobilealliance.org/

[13] http://www.iot-a.eu/public/public-documents/d3.1

[14] http://www.freescale.com/files/32bit/doc/white_paper/INTOTHNGSWP.pdf

[15] http://www.etsi.org/deliver/etsi_ts/102600_102699/102690/01.01.01_60/ts_102690v010101p.pdf

[16] http://docbox.etsi.org/M2M/Open/Latest_Drafts/00010ed211%20mIa,%20dIa%20and%20mId%20interfaces.pdf

[17] Constrained Application Protocol (CoAP)(http://tools.ietf.org/html/draftietf-core-coap-18).

Biographies

Soma Bandyopadhyay has more than 15 years of industry experience in the area of Embedded Systems, Digital Signal Processor, Protocol and Wireless Communications and ubiquitous computing. Since 2003 has been associated with Innovation lab of TATA Consultancy Services (TCS) as senior scientist. At present prime focus area for research is ubiquitous and sensor network computation, and Internet of Things. She has contributed towards the IEEE standard body on behalf of TCS. In TCS she has been involved in leading and development of 3GPP Long Term Evaluation (LTE/LTE-A) physical layer, IEEE-802.16d/e WMAN MAC stack development, Intel's graphics device driver. At present she is leading the research and development activity in the area of energy efficient M2M communication and its analytics. She worked on MPEG video decoder, dual core network processor target platform, Intel's network processor, protocol stack development on RTOS and embedded system, and multiple device drivers, ATM & MPLS.

Academically she is an M.Tech & B.Tech in Computer Science & Engineering from the University of Calcutta, India. She did her graduation in Physics (Hons.) from the same university.

 **Balamuralidhar P** is a Principal Scientist and Head of TCS Innovation Lab at Tata Consultancy Services Ltd (TCS), Bangalore. He has obtained Bachelor of Technology from Kerala University and Master of Technology (MTech) from IIT Kanpur. His PhD is from Aalborg University, Denmark in the area of Cognitive Wireless Networks. Major areas of current research include different aspects of Cyber Physical Systems, Sensor Informatics and Networked Embedded Systems. Before TCS his research careers were with Society for Applied Microwave Electronics Engineering & Research (SAMEER) Mumbai and Sasken Communications Ltd Bangalore.

He has over 25 years of research and development experience in Signal Processing, Embedded Systems and Wireless Communications. He has over 60 publications in various international journals and conferences and over 20 patent applications. Balamuralidhar was the leading TCS participation in two EU FP6 research consortium projects namely My Adaptive Global NET (MAGNET) and End to End Reconfigurability (E2R) in the area of next generation wireless communications. He is also contributing to TCS participation in National bodies like Broadband Wireless Consortium India (BWCI), Global ICT Standards for India (GISFI). In GISFI he is chairing the Internet of Things Workgroup.

Arpan Pal is a PhD from Aalborg University Denmark and is a senior member of IEEE. He had received his B.Tech and M.Tech from Indian Institute of Technology, Kharagpur, India in Electronics and Telecommunications.

He has more than 20 years of experience in the area of Signal Processing, Communication and Real-time Embedded Systems. Currently he is with Tata Consultancy Services (TCS), where he is heading research at Innovation Lab, Kolkata. He is also a member of Systems Research Council of TCS. His main responsibility is in conceptualizing and guiding R&D in the area of cyber-physical systems and ubiquitous computing with focus on applying the R&D outcome in the area Intelligent Infrastructure.

His current research interests include Mobile phone and Camera based Sensing and Analytics, Physiological Sensing, M2M communications and Internet-of-Things based Applications with focus on Energy, Healthcare and Transportation verticals. He has more than 40 publications till date in reputed Journals and Conferences along with a couple of Book Chapters. He has also filed for more than 35 patents and has 5 patents granted to him. He is an editor for IEEE Transactions on Emerging Topics in Computing for the special issue on Emerging Computing Technologies for Resilient & Robust Intelligent Infrastructure.

He had been earlier with Defense Research and Development Organization (DRDO) of Indian Govt. working on Missile Seeker Signal Processing. He has also worked with Macmet Interactive Technologies, leading their real-time systems group in the area of Interactive TV and Set-top boxes.